FLORA OF TROPICAL EAST AFRICA

RHAMNACEAE

MARSHALL C. JOHNSTON
(University of Texas Herbarium)

Trees, erect, climbing or scandent shrubs or lianes or subshrubs (or annual herbs but not in East Africa); tendrils present in *Helinus* and *Gouania*. Leaves simple, alternate or opposite, petiolate (or sessile but not in East Africa); blades penninerved or 3–5-nerved from the base, unlobed, serrate or crenate with each tooth or crenation usually associated with a minute gland, or entire. Stipules mostly present, free or interpetiolar or intra-axillary. Flowers minute, regular, bisexual (and often strongly protandrous, or reportedly protogynous in *Maesopsis*) or less commonly unisexual, perigynous or epigynous, 4- or usually 5-merous (6-merous very rarely and not in Africa), in basically cymose arrangements but the cymes often either reduced to fascicles (or even to solitary flowers) or arranged in short or elongate thyrses which in turn are sometimes disposed in leafy to leafless panicles; each flower with a cup lined with a thin intrastaminal nectariferous disk or the disk sometimes thickened near and/or produced beyond the rim of the cup and either free from the ovary or adnate to it. Sepals, petals and stamens attached at the rim of the cup. Sepals triangular, valvate in bud (this being one of the most useful traits to distinguish members of this family from those plants often confused with them). Petals absent or usually present, enclosed by calyx in bud, nearly always shorter than the sepals at anthesis, each usually with a narrow base or claw plus an expanded hood-like or concave or amplectant body closely associated with the stamen. Stamens bowed inward in bud, as many as, opposite to, usually shorter than and usually clasped or hooded by the petals. Ovary syncarpous, with 2 or 3 (rarely 4 or 1) cells; ovule solitary in the cell, anatropous; style minute, rarely simple, usually with 2, 3 or rarely 4 or even more rarely 10 microscopic stigmatic lobes at apex. Fruit often dryish and splitting into 3 1-seeded parts at maturity (as in the first 5 genera treated here), or fleshy and with 2 or 3 free 1-seeded stones (as in *Rhamnus* and *Scutia*) or fleshy or dryish and with a single 1-, 2- or 3-seeded (or 4-seeded but not in Africa) stone (as in the last 4 genera here); placentation basal. Seed with raphe dorsal or lateral; embryo large and straight, the cotyledons usually in planes tangential to the ovary-axis; endosperm in a thick or thin layer, rarely nearly absent but not in Africa, rarely ruminate but not in Africa.

A coherent natural family of perhaps 44 genera and 850 species, occurring in both tropical and temperate regions.

Several species are cultivated in East Africa, including *Colletia paradoxa* (Spreng.) Escal., a curious South American nearly leafless shrub with short, stout, laterally flattened, opposite thorns; *Ceanothus coeruleus* Lag., a well known Mexican blue-flowered ornamental shrub; and the eastern Asiatic *Hovenia dulcis* Thunb., a tree with curious swollen edible peduncles. Other cultivated species are mentioned under their respective genera.

Petioles longer than 2 mm.; leaf-blades usually longer than 2 cm. and not linear-lanceolate; stipules present though often caducous; plants not heath-like:
  Stipules not interpetiolar; leaves alternate, subopposite or opposite:
    Tendrils absent; ovary superior or inferior but even when inferior the fruit, through differential growth, more than half superior:
      Fruit either drupaceous or schizocarpous, not winged:
        Ovary and fruit wholly superior at all stages:
          Drupe 7–20 mm. long, or if longer then nearly globose and 2–3-celled; stigmatic area with 2 or 3 minute lobes; filaments well developed:
            Leaves alternate:
              Plants unarmed or with straight thorn-tipped branchlets; fruit when quite ripe with 2 or 3 free stones . . . . . . 6. **Rhamnus**
              Plants unarmed or with recurved spines at the nodes; fruit when quite ripe with a solitary stone 8. **Ziziphus**
            Leaves opposite or subopposite:
              Shrubs usually with short recurved thorns at the nodes; fruit globose or nearly so and when quite ripe with 2 or 3 free stones 7. **Scutia**
              Unarmed trees; fruits elongate with a single elongate stone . . . 9. **Berchemia**
          Drupe 20–30 mm. long, obovoid, with a single fertile locule; stigmatic area ± mushroom-like with about 10 minute marginal lobes; anthers subsessile . 11. **Maesopsis**
        Ovary inferior; fruit not wholly superior but fused tightly in the lower half with the large accrescent cup; unarmed weak or sprawling littoral shrubs . . . . 1. **Colubrina**
      Fruit dryish, produced apically into a strap-shaped wing several cm. long; strong lianes . . . . 10. **Ventilago**
    Coiled tendrils present; ovary and fruit wholly inferior:
      Leaves toothed; fruits at maturity, shortly before dehiscence, with 3 wings . . . 4. **Gouania**
      Leaves entire; fruit not winged even at full maturity . . . . 5. **Helinus**
  Stipules or most of them interpetiolar (at least on one side of each node); leaves or most of them opposite . . . . 2. **Lasiodiscus**
Petioles ± 2 mm. long; leaf-blades 1–2 cm. long, linear-lanceolate; stipules absent; somewhat heath-like shrubs with stiff ascending leafy branches . . . . 3. **Phylica**

1. COLUBRINA

Brongn., Mém. Fam. Rhamn.: 61 (1826); M.C. Johnst. in Wrightia 3: 91–96 (1963) & in Brittonia 23: 2–53 (1971), *nom. conserv.*

Macrorhamnus Baill. in Adansonia 11: 273 (1874)
Rhamnobrina H. Perr. in Not. Syst. Paris 11: 24 (1943)

Unarmed (or rarely armed but not in Africa) trees and shrubs, rarely scandent (as in the present species) but never a liane or twiner. Leaves alternate (or opposite but not in Africa). Stipules present, rarely (never in Africa) interpetiolar. Flowers bisexual, protandrous, 5-merous, in small axillary thyrses or fascicles. Disk massive and nearly completely surrounding the ovary. Ovary at anthesis inferior but later by differential growth becoming at least half superior. Cup and disk accrescent and persistently coherent to the 3(4)-celled ovary, at maturity covering the lower sixth to half of the fruit and at dehiscence breaking away ± irregularly along with the dryish (not spongy) exocarp, the 3 remnants adhering to the 3 crustaceous 1-seeded endo-mericarps which separate completely from each other, each releasing its seed through a longitudinal ventral slit.

A genus of 31 species, including 21 in the warmer parts of America, 1 in Hawaii, 4 in Madagascar, 4 in SE. Asia, and the present one.

Colubrina arborescens (Mill.) Sarg. (*C. ferruginosa* Brongn.), of tropical American forests, is a species of large shade tree planted here and there in East Africa (e.g., Tanganyika, Mwanza District, Ukerewe I., *Conrads* 717 !).

C. asiatica (*L.*) *Brongn.*, Mém. Fam. Rhamn.: 62 (1826); Hemsl. in F.T.A. 1: 383 (1868); T.T.C.L.: 466 (1949); Verdc. in B.J.B.B. 27: 359 (1957); K.T.S.: 388 (1961); R.B. Drummond in F.Z. 2: 430, t. 90 (1966). Type: Ceylon, *Hermann* 2: 11 (BM, lecto. !)

Clambering or sprawling evergreen woody vine or scandent shrub (or small tree *teste Semsei* 3053). Branchlets often zig-zag. Leaf-blades ovate, 3–10 cm. long, 1·5–6·3 cm. wide, broadly rounded at base to often shallowly cordate, acuminate, on each side with 15–34 rounded subappressed teeth, thin, glossy, green, nearly glabrous, beneath with a prominent midrib and 2 prominulent primary nerves and on each side with 2 remote secondary nerves; petioles 7–17 mm. long. Stipules deltoid, 1 mm. long. Inflorescences 5–10 mm. long; peduncle 2–4 mm. long; pedicels in flower 1–2 mm. long, in fruit to 5–15(–20) mm. long. Flowers protandrous, fragrant, rarely more than 2 of each cyme producing fruit. Cup ± 2 mm. broad at flowering. Style deeply 3-partite. Fruit nearly globose, 7–8 mm. thick and long (the cup about a fourth as long), promptly dehiscent, breaking first loculicidally at apex, thin-walled, the radial walls of the carpels membranous and occasionally parts of them adherent to the receptacle after dehiscence. Seed ventrally dihedral, dorsally convex, 4·5–6 mm. long, ± 4·5 mm. wide, of very low specific gravity.

var. **asiatica**

Younger internodes with sparse antrorse appressed golden silky hairs; leaf-blades usually 1·3–2 times as long as broad, at base broadly rounded, beneath with scattered antrorse appressed golden silky hairs, slightly more pubescent near the veins or often glabrate; seeds 4·5–5 mm. long. Fig. 1.

KENYA. Kwale District: 20 km. S. of Mombasa, Twiga, 22 Jan. 1964, *Verdcourt* 3963 !; Mombasa District: Likoni, Dec. 1957, *Ossent* 239 !; Kilifi District: 32 km. N. of Mombasa, Vipingo, 16 Dec. 1953, *Verdcourt* 1073 !
TANGANYIKA. Tanga, 4 Nov. 1929, *Greenway* 1834 !; Rufiji District: Mafia I., Kanga, 15 Aug. 1937, *Greenway* 5116 !; Lindi, 23 Mar. 1943, *Gillman* 1225 !

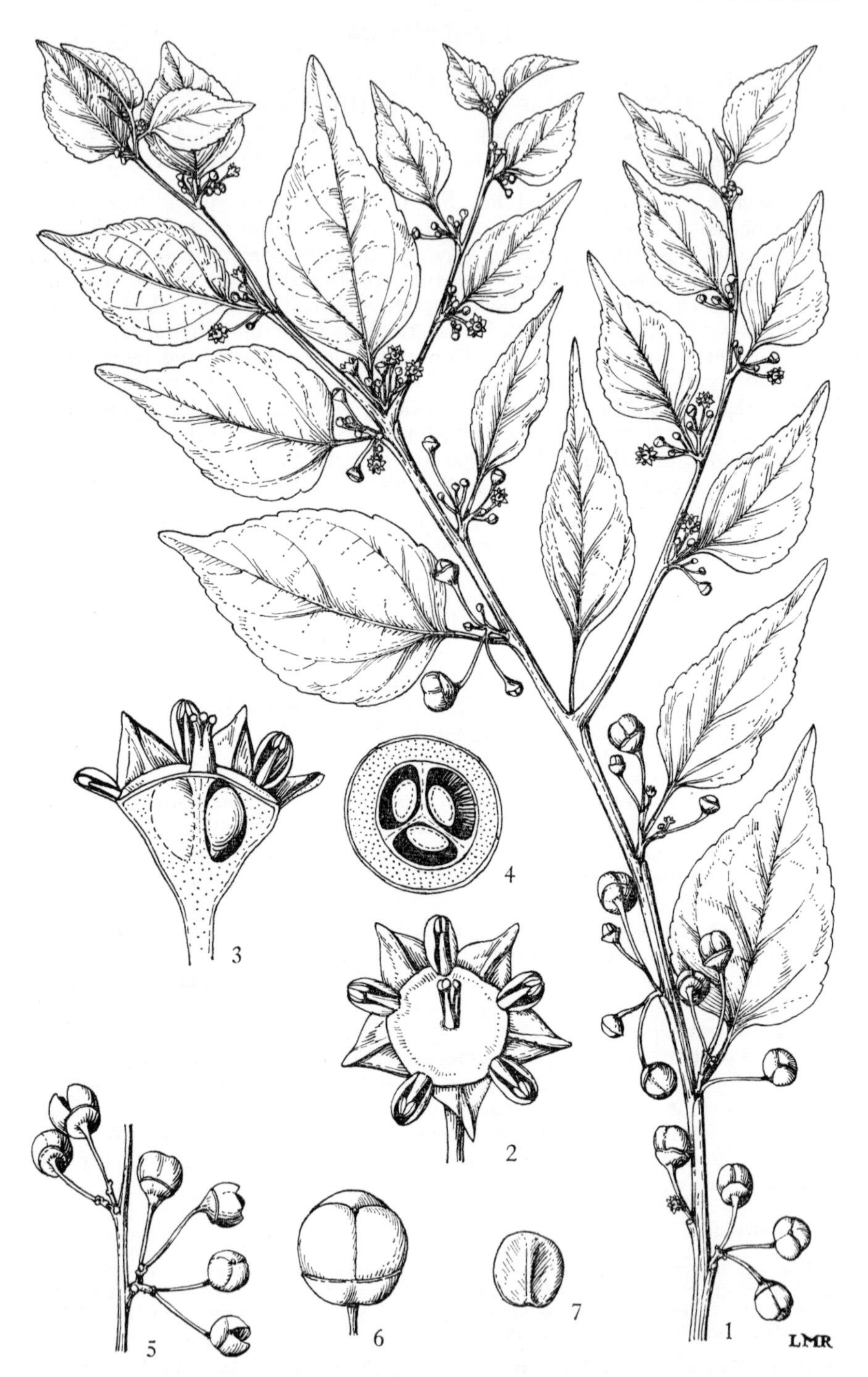

FIG. 1. *COLUBRINA ASIATICA* var. *ASIATICA*—**1,** branch with flowers and fruits, × $\frac{2}{3}$; **2,** flower, × 8; **3,** vertical section of flower, × 8; **4,** transverse section of ovary, × 8; **5,** portion of fruiting branch, × $\frac{2}{3}$; **6,** fruit, × 2; **7,** seed, × 2. 1–4, from *Drummond & Hemsley* 3252; 5–7, from *Volkens* 161. Reproduced from Flora Zambesiaca by permission of the Editors.

ZANZIBAR. Zanzibar I., Massazine, 13 Dec. 1959, *Faulkner* 2430!; Pemba I., Mkoani, 7 Sept. 1929, *Vaughan* 639!

DISTR. **K7**; **T3**, 6, 8; **Z**; **P**; tropical coasts, Kenya to Mozambique; Madagascar, Seychelles, SE. Asia, Queensland, and oceanic islands as far as 140° E. longitude in the S. Pacific; also Hawaii; also adventive to West Indies and Florida during the last century. The other variety is not littoral and occurs in Burma, Thailand, Laos, Cambodia, Vietnam, Yünnan and Java.

HAB. Just above high tide level

SYN. *Ceanothus asiaticus* L., Sp. Pl.: 196 (1753)

NOTE. According to H. B. Guppy, " Observations of a Naturalist in the Pacific 1896–99 " 2, Plant Dispersal (1906), the seeds float in sea-water; see also S. Carlquist in Brittonia 18: 310–335 (1966).

2. LASIODISCUS

Hook. f. in G.P. 1: 381 (1862)

Shrubs or usually trees. Leaves opposite or nearly so, unlobed, usually lanceolate-ovate, cuneate to rounded at base, usually acute, entire to serrate; petioles present but sometimes quite short. Stipules interpetiolar; at occasional nodes where the leaves are not quite opposite the stipules are not united, and on some branches the stipules are free on the upper side and interpetiolar on the lower. Inflorescence in the form of congested divaricate cymes or umbel-like or nearly reduced to axillary fascicles, usually rather long-pedunculate. Flowers perfect, protandrous. Cup usually broadly campanuloid. Sepals and petals 5. Disk massive and nearly completely surrounding the ovary. Ovary at anthesis inferior but later by differential growth becoming at least half superior; cup and disk accrescent and persistently coherent to the 3-celled ovary, at maturity covering the lower fourth or third of the fruit and at dehiscence breaking ± irregularly along with the dryish exocarp, the 3 remnants adhering to the 3 crustaceous 1-seeded meri-endocarps which separate elastically and completely from each other, each releasing its seed through a longitudinal ventral slit.

A tropical African and Madagascan genus of perhaps 12 species, in need of revision and only weakly distinguished from *Colubrina*.

Flowers in pedunculate divaricate cymes or umbel-like clusters or if the inflorescences short then the twigs pubescent:
  At least the younger parts hairy:
    Leaves auriculate or subcordate or usually shortly rounded at base; inflorescences many-flowered:
      Young shoots with hairs less than 1 mm. long . . . 1. *L. mildbraedii*
      Young shoots with hairs at least some of which are 1·1–2·3 mm. long . . . . 2. *L. ferrugineus*
    Leaves cuneate; cymes few-flowered (Uzaramo District, Pugu Hills) . . . . . 3. *L. holtzii*
  Plants glabrous (Zanzibar) . . . . . 5. *L. pervillei*
Flowers in few-flowered very shortly stalked umbel-like clusters; plants more or less glabrescent or sparsely hairy . . . . 4. *L. usambarensis*

1. **L. mildbraedii** *Engl.* in E.J. 40: 552 (1908); Burtt Davy & Bolton, Check-list Trees & Shrubs Uganda: 92 (1935); T.T.C.L.: 467 (1949); I.T.U., ed. 2: 323 (1952); Verdc. in B.J.B.B. 27: 362 (1957); R.B. Drummond in F.Z. 2: 438, t. 91/A (1966), pro parte, excl. syn. Type: Tanganyika, Bukoba District, Itara, *Mildbraed* 128 (B, holo.†)

Shrub or small tree 3–12(–24) m. tall. Older branches grey barked, lenticellate; young shoots finely but often densely pubescent with short spreading tawny hairs (some up to 0·5–0·8 mm. long). Leaf-blades elliptic to ovate, (5–)10–14(–19) cm. long, (2·3–)3–5(–8) cm. wide, firm, glabrous, asymmetric and very briefly rounded at base, acute or very slightly acuminate, glandular-crenulate or glandular-serrulate, on each side of midrib with 7–10 secondary nerves which are prominulent beneath; petioles only ± 3 mm. long. Stipules up to 13 mm. long, subulate, glabrescent, caducous. Inflorescences (5–)8–20-flowered, congested but divaricate-cymose, ± 2 cm. long and thick excluding peduncles; peduncles minutely and densely hairy, 2–6 cm. long; pedicels 3–7(–10) mm. long in flower, averaging somewhat longer (± 10–15, rarely to 20 mm.) in fruit, minutely and densely hairy. Flowers whitish. Cup densely pubescent externally with short whitish spreading hairs. Sepals ± 3 mm. long, widely spreading at anthesis. Petals oblanceolate, whitish, ± 2 mm. long, widely spreading at anthesis. Disk glabrous, green. Style ± 2 mm. long, 3-partite less than half the length. Fruit subglobose, ± 6–10 mm. thick, densely hairy with tawny sericeous hairs. Fig. 2.

UGANDA. Toro District: Bwamba, Nabulongwe Forest, 19 Dec. 1949, *Dawkins* 485!; Busoga District: Butembe-Bunya County, forest 1 km. N. of Lubanyi [Lubani] Hill, 7 Feb. 1951, *G. H. S. Wood* 59!; Masaka District: Malabigambo Forest, 6 km. SSW. of Katera, 2 Oct. 1953, *Drummond & Hemsley* 4541!

TANGANYIKA. Bukoba District: Minziro Forest Reserve, July 1950, *Watkins* 469! & Ihangiro, Sept.–Oct. 1935, *Gillman* 609!; Bagamoyo District: Kiona Plateau, Mar. 1966, *Mgaza* 828!

DISTR. **U**2–4; **T**1, 6; Mozambique

HAB. Lowland rain-forest and semi-swamp forest, usually forming an understorey in high forest; apparently below 1600 m.

SYN. *L. mildbraedii* Engl. var. *undulatus* Suesseng. in Mitt. Staatssamml. München 2: 40 (1954). Type: Mozambique, Manica e Sofala, Gorongosa Game Reserve, *Chase* 5084 (M, holo. !, BM, K, iso. !)

NOTE. Various authors have attributed *L. mildbraedii* to W. Africa, Congo and Sudan, apparently on the basis of misdetermined specimens of *L. mannii* Hook. f.; cf. N. Hallé in Adansonia, n.s., 2: 129–133 (1962).

The *Mgaza* collection cited above appears to be correctly assigned although the plant is described as a " semi shrub up to 4 ft. tall ". It represents a considerable extension of the previously known distribution.

The wood is durable and has a variety of local uses.

2. **L. ferrugineus** *Verdc.* in B.J.B.B. 27: 362 (1957); K.T.S.: 388 (1961). Type: Kenya, Lamu District, Utwani Forest, *Rawlins* 241 bis (EA, holo. !, BR, K, iso. !)

Much-branched evergreen shrub or small tree to 6 m. tall. Several-year-old branches with pale grey bark; young twigs densely ferrugineous tomentose, with some spreading silky hairs 1·1–2·3 mm. long as well as a shorter denser pubescence. Leaf-blades elliptic to elliptic-oblong, 6–18·5 cm. long, 2·2–6·5 cm. wide, at base rounded and minutely subcordate, acuminate, acute, shortly serrulate, glabrous above, beneath on the nerves tawny pilose, on each side of midrib with 8–10 secondary nerves which are prominulent beneath; petioles 5–8 mm. long, hirsute. Stipules lanceolate, 7–12 mm. long, pubescent. Cymes congested-dichotomous, 15–40-flowered; peduncles 4–7 cm. long, tomentose and pilose; secondary peduncles up to 22 mm. long; pedicels tomentose, 9–12 mm. long in flower. Cup densely hirsute or villous externally. Sepals 3–3·5 mm. long, externally villous. Petals linear-obovate, ± 2 mm. long. Style 3-partite about half the length. Fruit unknown.

KENYA. Kilifi District: near Malindi, Kikuyuni, 26 Dec. 1954, *Verdcourt* 1181!; Lamu District: NE. of Witu, 28 Feb. 1956, *Greenway & Rawlins* 8954! & Utwani Forest, Oct. 1937, *Dale* in *F.D.* 3834!

FIG. 2. *LASIODISCUS MILDBRAEDII*—**1,** flowering branch, × 1; **2,** leaf, showing undersurface, × 1; **3,** flower bud, × 4; **4, 5,** flower, viewed from side and above respectively, × 4; **6,** petal, × 8; **7,** stamens, front and back views, × 8; **8,** gynoecium, × 4; **9,** transverse section of ovary, × 4; **10,** ovule, × 2; **11,** fruit, × 1; **12,** portion of dehisced fruit, × 1. All from *Greenway & Eggeling* 7349.

DISTR. **K7**; not known elsewhere
HAB. Lowland evergreen forest; less than 100 m.

NOTE. Drummond in F.Z. 2: 438 (1966) has submerged this in *L. mildbraedii*. The relationship is indeed close, but the present plant deserves recognition at some taxonomic level. Perhaps a future monographer will see fit to reduce it to varietal status.

3. **L. holtzii** *Engl.* in E.J. 40: 551 (1908); T.T.C.L.: 467 (1949). Type: Tanganyika, Uzaramo District, Pugu Hills, near Kiserawe, *Holtz* 935 (B, holo. †)

Branching shrub. Older branchlets fuscous, glabrescent; twigs densely ferrugineous tomentulose. Leaf-blades oblong, 8–11 cm. long, 3–4 cm. wide, cuneate, acuminate, subcoriaceous, glabrous except the principal nerves beneath pilose, serrulate, the primary and secondary nerves beneath prominulent; petioles ± 5 mm. long, ventrally appressed pilose. Stipules lanceolate, ± 5 mm. long, caducous, densely appressed pilose. Inflorescences few-flowered; peduncles short, densely tomentulose; pedicels 5 mm. long. Cup pubescent externally. Sepals ± 2 mm. long, pubescent dorsally. Petals ± 1·5 mm. long, obovate. Disk 2–2·5 mm. wide. Style 0·5 mm. long at an early stage (in anthesis). Fruit unknown.

TANGANYIKA. Uzaramo District: Pugu Hills, near Kiserawe, Apr. 1903, *Holtz* 935
DISTR. **T6**; known only from the type
HAB. Apparently woodland; 100–200 m.

NOTE. Apparently the species has been collected only once. No modern specimen matches the description.

4. **L. usambarensis** *Engl.* in E.J. 40: 551 (1908); T.T.C.L.: 467 (1949); Verdc. in B.J.B.B. 27: 361 (1957); R.B. Drummond in F.Z. 2: 438 (1966). Type: Tanganyika, Lushoto District, E. Usambara Mts., Derema [Nderema], *Scheffler* 186 (B, holo. †, EA, K, iso. !)

Shrubs or trees 2–6 m. tall. Older twigs glabrate, fuscous; younger twigs sparsely and shortly pilose. Leaf-blades narrowly elliptic to oblong-lanceolate (rarely nearly ovate), 9–14(–17) cm. long, 3–5(–6) cm. wide, cuneate, acute or obtuse, acuminate, crenate-serrulate to sharply serrulate, firm, glabrous except on the major nerves beneath, on each side of midrib with 7–9 secondary nerves which are prominulent beneath; petioles 3–9 mm. long, pubescent ventrally. Stipules 5–9 mm. long, subulate, pubescent, caducous. Inflorescences short, with (3–)6–15 flowers, umbel-like; peduncles 3–4 mm. long; pedicels 3–7 mm. long at anthesis, later 10–15 mm. long. Cup externally minutely pubescent. Sepals ± 2 mm. long, dorsally minutely pubescent. Petals broadly oblanceolate, 1·2–1·5 mm. long, whitish. Disk ± 2 mm. broad. Style ± 1 mm. long, 3-partite about half the length. Fruit nearly globose, 8–9 mm. thick, minutely and densely pubescent.

var. **usambarensis**

Peduncles, pedicels and sepals merely pubescent, not velutinous. Leaves acute.

TANGANYIKA. Morogoro District: Uluguru Mts., Bondwa Peak above Morningside, Jan. 1953, *Eggeling* 6452 ! & Uluguru Mts., NW. side, 8 Nov. 1932, *Schlieben* 2924 !; Ulanga District: SSW. of Mahenge, Muhulu Mt., 30 Jan. 1932, *Schlieben* 1683 !
DISTR. **T3, 6**; also Rhodesia *fide* Drummond in F.Z. 2: 439; the var. *gossweileri* Cavaco occurs in Angola
HAB. Apparently mist-forest and rain-forest; 900–1650 m.

NOTE. This is exceedingly closely related to the West African *L. fasciculiflorus* Engl. Further work, especially field work, is needed to clarify whether this is worthy of distinction at a specific level.

5. **L. pervillei** *Baill.* in Adansonia 8: 202 (1868); Perr. in Not. Syst. Paris 11: 26 (1943) & in Fl. Madag. & Com., fam. 123: 29 (1950); Capuron in Adansonia, n.s., 6: 136 (1966). Type: Madagascar, Nossi Mitsio I. [Nossy-Mitsieu], Sambirano, *Pervillé* 321 (P, holo.!)

Shrub about 2–3 m. tall or small tree to 8–10 m. tall. Older branchlets greyish, lenticellate; youngest twigs as well as the rest of the herbage glabrous. Leaves deciduous at certain seasons; blades lanceolate-ovate or elliptic, (5–)9–15(–20) cm. long, (1·7–)3–6(–8) cm. wide, at base rounded or obtuse and usually slightly asymmetric, attenuate-obtuse at apex, serrulate, on each side of midrib with 7–8 secondary nerves which are prominulent beneath; petioles 3–8 mm. long. Stipules very firm, lanceolate-subulate, ± 7 (rarely to 16) mm. long and 4·5 mm. broad, deciduous. Inflorescence cymose, branched 2 or 3 times at base of cyme, with (4–)8–12(–20) flowers; peduncles (7–)25–50(–80) mm. long; pedicels 1–2 cm. long at anthesis, up to 3 cm. long in fruit. Cup minutely puberulent, glabrescent. Sepals ± 3 mm. long, reflexed at anthesis. Petals almost linear, slightly shorter than the sepals, caducous. Disk 2 mm. across, thick, glabrous. Style 3-lobed ± half the length. Fruit roughly globose, ± 8 mm. long. Seed orbicular, the ventral dihedral nearly plane, 6–7 mm. long.

ZANZIBAR. Zanzibar I., without precise locality, *Sacleux* 1076! & Haitajwa Hill, 28 Jan. 1929, *Greenway* 1206! & 20 Sept. 1930, *Vaughan* 1566!
DISTR. **Z**; Madagascar

HAB. Unknown (in Madagascar widespread in forests and woodlands)

3. PHYLICA

L., Sp. Pl.: 195 (1753) & Gen. Pl., ed. 5: 90 (1754): Pillans in Journ. S. Afr. Bot. 8: 1–464 (1942)

Shrubs (or small trees but not in East Africa), often in parts densely pubescent, often ericoid, with stiff erect branches and numerous stiff crowded leaves. Leaves alternate, shortly petiolate (to nearly sessile but not in East Africa); blade-margins usually strongly revolute and entire or if toothed the toothing rolled out of sight. Stipules absent (present in a South African species). Flowers bisexual, regular, 5-merous, in tight head-like clusters (contracted thyrses) or in more expanded sometimes raceme-like thyrses (but not in East Africa), the heads often with extremely pubescent bracts. Cup variously bell-shaped or cylindrical, either only slightly or much surpassing the ovary. Sepals 5. Petals 5 (or absent but not in East Africa), cucullate or merely concave. Stamens 5. Disk very variable, usually thickened around the top of the ovary where it is free from the cup. Ovary inferior at all stages, 3-celled, often top-shaped, less commonly cylindrical; style simple with 3 minute stigmatic lobes. Fruit a dryish schizocarpous capsule, tardily separating into the 3 mericarps each of which liberates its seed through a ventral cleft. Seed with a lobed structure embracing the base, this often called an " aril ".

A genus of about 100 species (Pillans claims 150), principally South African but occurring north as far as southern Tanganyika, also on Madagascar and various other islands (St. Helena, Tristan da Cunha group, Gough, New Amsterdam, Réunion).

P. emirnensis (*Tul.*) *Pillans* in Journ. S. Afr. Bot. 8: 26 (1942). Type: Madagascar, Emirna region, *Bojer* (P, holo., M, iso.!)

Small shrubs up to 4 dm. tall (or up to 1·2 m. *fide Richards* 7581) with numerous stems from the base, these branched only sparingly (at the bases of old inflorescences), the branches also stiffly erect, brownish with a dense pubescence of greyish antrorse hairs about 0·2 mm. long. Leaf-blades

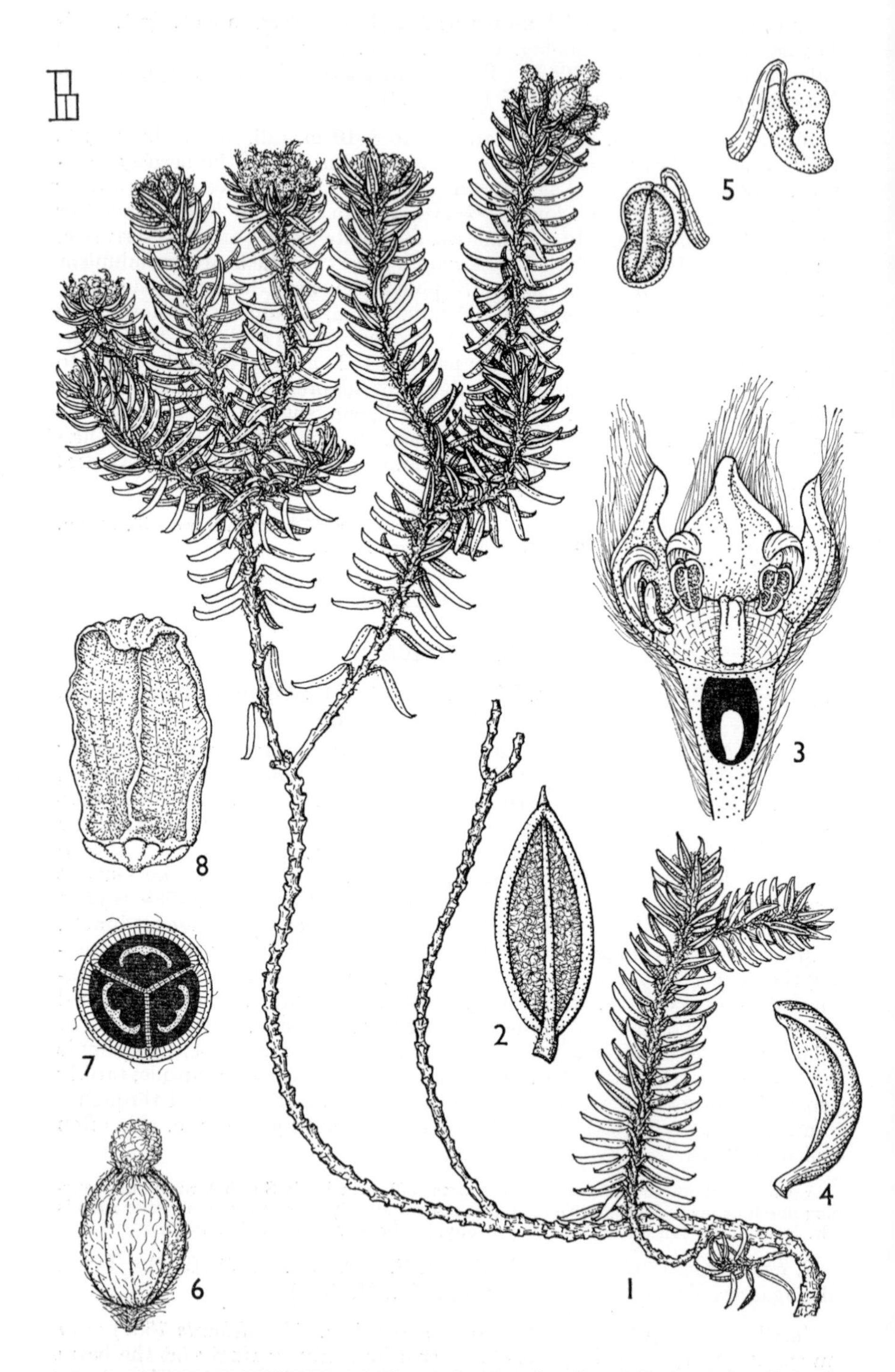

FIG. 3. *PHYLICA EMIRNENSIS*—**1,** habit, × 1; **2,** undersurface of unrolled leaf, × 4; **3,** vertical section of flower, × 10; **4,** petal, × 40; **5,** stamens, front and back views, × 20; **6,** fruit, × 3; **7,** transverse section of fruit, × 4; **8,** seed, × 10. 1–5, from *Eggeling* 6607; 6–8, from *Richards* 7581.

lanceolate, 6–8(–10, or –17 *fide* one author) mm. long, 3–4 mm. wide (but because of the rolling back of the margins appearing only 2–3 mm. wide), acute and mucronulate, very firm to coriaceous, above dark, glabrous, shiny and somewhat sculpted near the margins, beneath white tomentulose; petiole ± 1 mm. long. Heads terminal, 3–7-flowered, ± 1 cm. thick at anthesis, ± 2 cm. thick in fruit. Bracts densely white-woolly and about as long as the fruit. Flowers 4–5 mm. long including the hairs. Cup and sepals uniformly externally pubescent with white hairs ± 1·5 mm. long (or the cup less pubescent in the Madagascar populations). Cup elongated ± 1 mm. beyond the ovary. Sepals ± 1 mm. long. Petals oblanceolate, ± 0·4 mm. long, nearly glabrous. Fruit 6–7 mm. long, 4–5 mm. thick, sparsely villous, pinkish when not quite mature. Fig. 3.

TANGANYIKA. Njombe District: Elton Plateau, May 1953, *Eggeling* 6607 ! & Kipengere, top of mountain, 7 Jan. 1957, *Richards* 7581 ! & Elton Plateau, Dec. 1959, *Zebedayo* 31 !

DISTR. **T7**; Madagascar (Centre)

HAB. Rock outcrops on mountain tops; 2400–2600 m.

SYN. *Tylanthus emirnensis* Tul. in Ann. Sci. Nat., sér. 4, 8: 128 (1857)
[*Phylica tropica* sensu Engl. in E.J. 30: 351 (1902); R.B. Drummond in F.Z. 2: 432 (1966), pro parte saltem quoad specim. Tanganyik., *non* Bak.]
P. emirnensis (Tul.) Pillans var. *nyasae* Pillans in Journ. S. Afr. Bot. 8: 27 (1942), *nom. invalid., sine lat. diagn.*; T.T.C.L.: 468 (1949); Verdc. in B.J.B.B. 27: 359 (1957). Type: Tanganyika, Njombe District, Kinga Mts., Kipengere ridge, *Goetze* 963 (B, holo.†, BR, iso. !)

NOTE. The collections of *Phylica* from the high mountains of tropical Africa and Madagascar resemble each other so closely that one wonders whether *P. emirnensis* and *P. tropica* Bak. in K.B. 1898: 302 should be considered conspecific. The distinctions given by Pillans in Journ. S. Afr. Bot. 8: 28 (1942) are highly technical and perhaps entirely trivial. The collections are so few that a critical revision is scarcely feasible at present.

4. GOUANIA

Jacq., Select. Stirp. Amer. Hist.: 263 (1763)

Climbing shrubs or lianes with circinnate tendrils (rarely, or apparently never in Africa, seeming to be mere shrubs). Leaves alternate, petiolate; blades usually ovate, serrate (rarely nearly entire), acute or acuminate, at base usually rounded or cordate and often 3–5-nerved but penninerved more distally. Stipules present, free, usually caducous. Cymes usually small, but usually aggregated into ample leafless terminal panicles. Flowers 5-merous, bisexual, epigynous at all stages. Disk rather massive, epigynous. Ovary usually urceolate; style trifid or obscurely 3-lobed. Fruit a dry schizocarpous capsule, late in maturation the endo- and exocarpous tissue near the septa (on both sides of each mericarp) producing 3 broad thin wings (thus each wing is composed of parts of 2 adjacent mericarps; reportedly the wing-production fails in certain Asian and American, but not African, populations, and perhaps the reports are based on observations of immature fruit, for the wing-proliferation is a quite late development).

A pantropical genus with roughly 15 species in the Americas, perhaps 4 in Asia, about 5 in Madagascar and other islands of the Indian Ocean, and another 2 in Africa. It is difficult to be precise regarding the numbers of species because the genus is so badly in need of revision.

Leaves at maturity only slightly pubescent . . 1. *G. scandens*
Leaves at maturity densely pubescent beneath . 2. *G. longispicata*

1. **G. scandens** (*Gaertn.*) *R.B. Drummond* in F.Z. 2: 435, t. 88/D (1966). Type: Mauritius [Isle de France], *Hermann* (?TUB, not located there, apparently lost; lecto.: Gaertner's protolog cited below)

Liane or climbing or sprawling shrub several m. long; branchlets and tendrils with longitudinal densely pubescent stripes of antrorse brown hairs 0·2–0·3 mm. long, glabrescent, very slightly zig-zag. Leaf-blades broadly ovate, 4–8 cm. long, 2·4–5·8 cm. wide, very shallowly cordate to rounded, acuminate (the acumen 5–10 mm. long), serrulate or doubly serrulate, when young with dense pubescence near the veins of brownish antrorse hairs 0·2–0·4 mm. long, this pubescence much sparser and less noticeable on the mature blade, on each side of midrib with 6–8 secondary nerves; petioles 8–25 mm. long, ventrally densely pubescent with brownish hairs 0·3–0·5 mm. long. Stipules linear, ± 3 mm. long, caducous. Thyrses merely terminal, not aggregated into panicles, 4–8 cm. long, lax, ± 1 cm. thick in flower, axes densely pubescent at least when young, cymes few-flowered; peduncles of cymes ± 1 mm. long, pubescent; pedicels 0·5–1 mm. long, pubescent. Flowers yellowish or yellowish white. Cup ± 2 mm. wide, pubescent externally. Sepals ± 0·9 mm. long. Petals ± 0·9 mm. long. Style 0·3–0·5 mm. long or perhaps longer in later stages of anthesis (only young flowers seen). Fruit said to be 1·3–1·7 mm. long and broad with prominently reticulate wings.

TANGANYIKA. Lindi District, Lake Lutamba, 21 Nov. 1934, *Schlieben* 5645! & 20 km. S. of Lindi, Mlinguru, 18 Dec. 1934, *Schlieben* 5733!
DISTR. T8; Mozambique; Mauritius and other islands of the Indian Ocean
HAB. Apparently near lakes and rivers; 200–275 m.

SYN. *Retinaria scandens* Gaertn., Fruct. 2: 187, t. 120/4 (1791)
Gouania retinaria DC., Prodr. 2: 40 (1825), *nom. superfl., illegit.*
[*G. longipetala* sensu Hemsl. in F.T.A. 1: 383 (1863), pro parte quoad specim. Kirk, *non* Hemsl. sensu stricto]
[*G. tiliifolia* sensu Bak., Fl. Maurit. & Seychelles: 52 (1877), pro parte; T.T.C.L.: 467 (1949); Suesseng. in E. & P. Pf., ed. 2, 20d: 168 (1953), pro parte quoad syn. *G. retinaria, non* Lam.]
G. mozambicensis M.L. Green in K.B. 1916: 199 (1916); Suesseng. in E. & P. Pf., ed. 2, 20d: 168 (1953). Type: Mozambique, Chupanga, 1860, *Kirk* (K, holo. !)

NOTE. The application of Gaertner's name is not entirely clear, but until a critical revision can be undertaken the treatment of R.B. Drummond in F.Z. is followed here.

2. **G. longispicata** *Engl.*, P.O.A. C: 256 (1895); Bak. f. in J.L.S. 40: 45 (1911); Eyles in Trans. Roy. Soc. S. Afr. 5: 407 (1916); M.L. Green in K.B. 1916: 198 (1916); Brenan, T.T.C.L.: 466 (1949) & in Mem. N.Y. Bot. Gard. 8: 239 (1953); Suesseng. in E. & P. Pf., ed. 2, 20d: 168 (1953); Verdc. in B.J.B.B. 27: 359 (1957); F.W.T.A., ed. 2, 1: 670 (1958); Evrard in F.C.B. 9: 449 (1960); E.P.A.: 502 (1960); R.B. Drummond in F.Z. 2: 434, t. 88/C (1966). Type: Congo, Undussuma, *Stuhlmann* (B, holo. †)

Climbing or sprawling shrubs or lianes several m. long. Branchlets and tendrils with longitudinal pubescent stripes of antrorse brownish hairs 0·2–0·3 mm. long, glabrescent, very slightly zig-zag. Leaf-blades ovate, 4–8·5 cm. long, 2·5–4·5(–7) cm. wide, round at base, acute or acuminate (acumen 5–10 mm. long), serrulate or doubly serrulate, above dark and with sparse antrorse golden-brown hairs 0·2–0·4 mm. long, beneath with a dense persistent grey or brown tomentum, on each side of midrib with 6–7 secondary nerves; petioles 10–25 mm. long, ventrally pubescent. Stipules subulate, ± 3 mm. long (rarely to 10 mm.). Thyrses merely terminal or aggregated into loose panicles; cymes 3–10-flowered; peduncles of cymes 1 mm. long and pubescent or absent; pedicels 2–3 mm. long, pubescent. Flowers whitish or greenish white, fragrant and nectariferous. Cup ± 2 mm. wide, externally pubescent. Sepals ± 1 mm. long. Petals ± 1 mm. long. Style ± 0·5 mm. long. Fruit 6–8 mm. long, 8–11 mm. wide. Fig. 4.

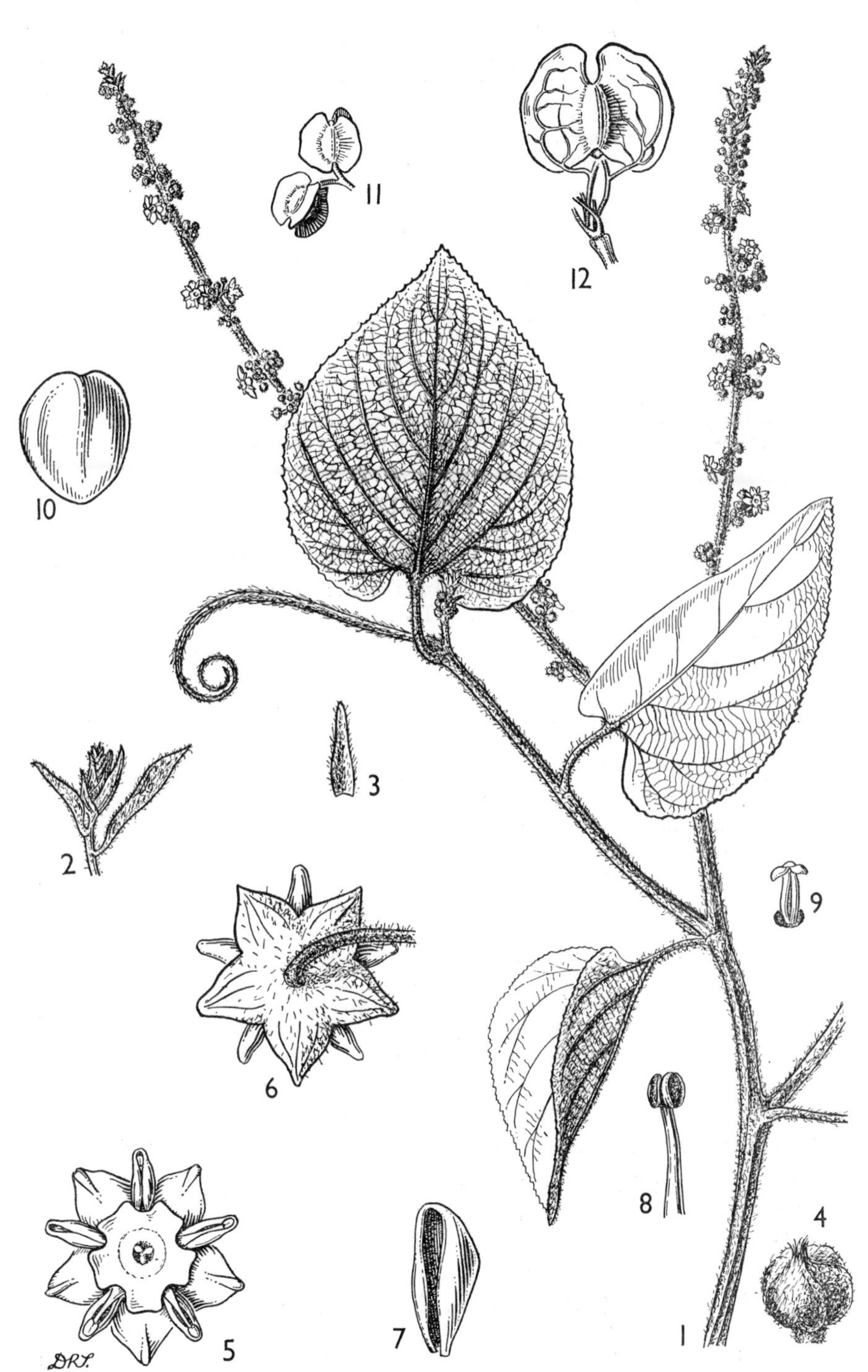

Fig. 4. *GOUANIA LONGISPICATA*—**1,** flowering branch, × 1; **2,** tip of young shoot, × 1; **3,** bract, × 2; **4,** flower bud, × 6; **5, 6,** flower, viewed from above and beneath respectively, × 6; **7,** petal, × 12; **8,** stamen, × 12; **9,** style, × 8; **10,** ovule, × 6; **11,** fruit, × 1; **12,** portion of dehisced fruit, × 2. All from *Purseglove* 726.

UGANDA. Kigezi District: Kinkizi, Amahenge, Nov. 1946, *Purseglove* 2249!; Busoga District: 16 km. SE. of Bugiri, Igwe, 26 May 1951, *G.H.S. Wood* 240!; Mengo District: Bujuko, Lwamunda Forest, 6 June 1951, *Dawkins* 754!

KENYA. Kiambu District: Karura Forest, 21 Nov. 1966, *Perdue & Kibuwa* 8044!; Kericho District: SW. Mau Forest, Sambret Catchment, 16 Aug. 1962, *Kerfoot* 4079!; Teita Hills, Ngangao Forest, *Gardner* in *F.D.* 2979!

TANGANYIKA. Moshi District: Lyamungu, 23 Aug. 1932, *Greenway* 3122!; Lushoto District: Shagayu Forest, 6 Oct. 1964, *Mgaza* 632!; Mpanda District: Kasangazi, 31 July 1958, *Mahinde* 187!

DISTR. **U**1–4; **K**3–5, 7; **T**2–4, 6–8; Nigeria, Congo, Sudan, Rhodesia, Malawi, Mozambique

HAB. Forests, particularly at margins and in disturbed places, riverine thickets and wooded grassland; 300–2400 m.

NOTE. *G. ulugurica* Gilli in Ann. Nat. Mus. Wien 74: 444, t. 5/1 (1970) is based on a sterile specimen, *Gilli* 321 (W, holo.) from Tanganyika, Morogoro District, Chenzema [Tchenzema]. It lacks the characteristic persistent woolly tomentum on the leaf-undersurface, the indumentum restricted to sparse golden brown hairs principally on the main nerves. *Greenway* 10477 another sterile specimen from Lushoto District, Mkusi, is similar. Without fertile parts the status of these plants is uncertain, but quite probably they represent no more than a minor variant.

5. HELINUS

Endl., Gen.: 1102 (1840), *nom. conserv.*

Mystacinus Raf., Sylv. Tellur.: 30 (1838)
Marlothia Engl. in E.J. 10: 39 (1888)

Climbing shrubs or woody vines or lianes (or seeming to be mere shrubs but not in East Africa). Tendrils present (or absent but not in East Africa), circinnate, axillary. Leaves alternate, petiolate, the uppermost ones (subtending umbels) sometimes much reduced and bract-like; blades ovate or lanceolate, entire, at base rounded or cordate, at apex acute or rounded, essentially penninerved or vaguely 3–5-nerved at base. Stipules linear, caducous. Umbels axillary, (1–)3–12-flowered, only 1 flower or rarely 2 flowers per umbel setting fruit; peduncles usually rather slender and elongate, sometimes surpassing the subtending leaf; pedicels short to elongate, filiform. Flowers 5-merous, perfect, epigynous at all stages. Cup broadly campanulate. Sepals deltoid, often with very thin white margins. Petals oblanceolate and somewhat concave toward the distal end, yellowish or greenish white, arcuate ascending at anthesis. Stamens as long as petals. Disk massive, thickened around and adherent to the ovary. Ovary inferior at all stages of development, 3-celled (reportedly sometimes 2-celled but not in East Africa). Style 3-lobed about half the length. Fruit a schizocarpous capsule with a thinly coriaceous rind (mesocarp and other exterior tissues) which eventually dries; endocarp thin, crustaceous, schizocarpous, separating into 3 mericarps each of which releases its seed by a sudden opening of the ventral suture.

A genus of 5 species, one in north-western India, one in Madagascar and 3 in Africa.

Flowers glabrous; fruits smooth and glabrous . . 1. *H. integrifolius*
Flowers externally pubescent; fruit glandular-tuberculate, the tubercles crowded and acute and with spreading hairs . . . . 2. *H. mystacinus*

1. **H. integrifolius** (*Lam.*) *Kuntze*, Rev. Gen. 1: 120 (1891); Exell & Mendonça, C.F.A. 2: 31, t. 6 (1954); Evrard in F.C.B. 9: 450 (1960); F.F.N.R.: 434 (1962); R.B. Drummond in F.Z. 2: 436, t. 92/A (1966). Type: a sterile specimen cultivated in Paris, collector unknown (P-LA, holo.!)

Woody climber to 6 m. or more. Herbage usually only slightly if at all pubescent, sparsely appressed pilose especially on the younger shoots and the nerves of the young blades. Leaf-blades ovate to broadly ovate, (1–)3–6 cm. long, (7–)17–50 mm. wide, at base rounded or usually shallowly cordate, at apex rounded or rarely acutish, mucronate, thin, at base 3-nerved, above the base on each side of midrib with 3–6 secondary nerves; petioles filiform, (3–) 7–28 mm. long. Stipules linear, (2–)3–6 mm. long, caducous. Peduncles filiform, (1–)2–5 cm. long; pedicels filiform, (3–)5–8 mm. long in flower, (5–) 7–10 mm. long in fruit. Cup glabrous. Sepals ± 2 mm. long, glabrous. Petals ± 1·8 mm. long, whitish. Fruit pendulous, globose to slightly obovoid-globose, 5–7 mm. long, glabrous, ripening through reddish green to dark brown or black.

KENYA. Machakos District: Mtito Andei, 25 Dec. 1945, *Bally* 4719!; Masai District: 61 km. on Magadi–Nairobi road, Olenyamu, 30 June 1962, *Glover, Gwynne & Samuel* [*Pavlo*] 2927!; Teita District: Tsavo National Park East, Voi Gate Camp Site, 21 Dec. 1966, *Greenway & Kanuri* 12816!

TANGANYIKA. Mbulu District: Lake Manyara National Park near the main gate, 28 Feb. 1964, *Greenway & Kanuri* 11260 in part, mixed with *Helinus mystacinus*!; Dodoma District: 80 km. on Dodoma–Iringa road, 18 Feb. 1932, *St. Clair-Thompson* 375!; Morogoro District: Turiani, 13 July 1936, *Rounce* 469!

DISTR. **K**1, 2, 4, 6, 7; **T**1–7; Yemen, Socotra, Somali Republic, Congo, Angola, Malawi, Rhodesia, Mozambique, South West Africa, South Africa

HAB. Thickets in wooded grassland, forest margins, bushland and dry open woody vegetation of all sorts; 0–1700 m.

SYN. *Gouania integrifolia* Lam., Encycl. 3: 5 (Oct. 1789)
Willemetia scandens Eckl. & Zeyh., Enum. Pl. Afr. Austr. Extratrop. 1: 130 (1834). Types: South Africa, Cape Province, Uitenhage, Zuurberg Mts., near Enon, & Kat R. near Balfour and Philipstown, *Ecklon & Zeyher* 996 (M, isosyn.!)
Helinus scandens (Eckl. & Zeyh.) A. Rich., Tent. Fl. Abyss. 1: 139 (1847); Staner in B.J.B.B. 15: 415 (1939); T.T.C.L.: 467 (1949); Suesseng. in E. & P. Pf., ed. 2, 20d: 172 (1953); Verdc. in B.J.B.B. 27: 360 (1957); F.F.N.R.: 227 (1962)
H. arabicus Jaub. & Spach, Illustr. 5: 81, t. 472 (1856). Type: Yemen, Mt. Maammara, *Botta* (P, holo.!)
H. ovatus Sond. in Fl. Cap. 1: 479 (1860), as "*ovata*"; Hemsl. in F.T.A. 1: 384 (1868), *nom. illegit.*, based in part on *Willemetia scandens* Eckl. & Zeyh.
H. ovatus Sond. var. *rotundifolius* Sond. in Fl. Cap. 1: 479 (1860), as "*rotundifolia*". Type: South Africa, Durban, [Port Natal], *Gueinzius* 399 (? TCD, iso.)
Mystacinus arabicus (Jaub. & Spach) Kuntze, Rev. Gen. 3(2): 39 (1898)
?*Helinus scandens* (Eckl. & Zeyh.) A. Rich. var. *parvifolius* Engl., V.E. 3(2): 316 (1921). Type: Ethiopia/Somali Republic, Boran, collector not stated (B, holo.†)

NOTE. The application of Lamarck's name to the smooth-fruited plants is simply an arbitrary action on the part of Kuntze and of Exell & Mendonça and more recent authors of African floras. In the absent of flowers and fruits, one cannot exclude the possibility that Lamarck's plant and Aiton's *Rhamnus mystacinus* (below) are conspecific. The degree of herbage-pubescence, especially in cultivated plants, cannot be considered diagnostic. Furthermore, one may even question whether the smooth-fruited and rough-fruited plants are specifically distinct. The smooth plants range farther west in Africa (to Angola) and have not been found in Ethiopia, but otherwise the distributions are nearly co-extensive. The overall similarities in the plants are quite overwhelming, and the technicality of fruit-texture may be under a simple genetic control. Progeny tests in areas where the two sorts of plants occur in proximity should establish quickly whether both smooth-fruited and rough-fruited plants may be obtained from the seed of a single plant. Aiton's name is earlier than Lamarck's.

2. **H. mystacinus** (*Ait.*) *Steud.*, Nom., ed. 2, 1: 742 (1840); Hemsl. in F.T.A. 1: 385 (1868); Bak. f. in J.L.S. 40: 45 (1911); Eyles in Trans. Roy. Soc. S. Afr. 5: 407 (1916); Staner in B.J.B.B. 15: 417 (1939); T.T.C.L.: 467 (1949); Suesseng. in E. & P. Pf., ed. 2, 20d: 172 (1953); Verdc. in B.J.B.B. 27: 360 (1957); Evrard in F.C.B. 9: 452, t. 46 (1960); E.P.A.: 502 (1960); F.F.N.R.: 227, t. 39 (1962); R.B. Drummond in F.Z. 2: 436, t. 92/B (1966). Type: a

FIG. 5. *HELINUS MYSTACINUS*—**A, B,** branches with flowers and fruits respectively, × ½; **C, D,** short- and long-styled flowers respectively, × 5; **E,** vertical section of long-styled flower, × 5; **F,** petal, × 10; **G,** stamen × 10; **H,** transverse section of ovary, × 10; **I,** fruit, × 2. All from *de Witte* 1118. Reproduced by permission of the Director of the Institut National pour l'Étude Agronomique du Congo.

plant cultivated in 1775 at Kew from seed sent by Bruce probably from NE. Africa (BM-Banks, holo., TEX, photo. !)

Woody climbers to 10 m. or more. Herbage usually appressed pilose to nearly tomentulose but occasionally nearly glabrous, the youngest shoots and nerves of the young blades being the most pubescent parts. Leaf-blades ovate to broadly ovate, (1–)3–6 cm. long, (7–)17–50 mm. wide, at base rounded or usually shallowly cordate, at apex rounded or rarely acutish, mucronate, thin, at base 3-nerved, above the base on each side of midrib with 3–6 secondary nerves; petioles filiform, (3–)7–20 mm. long. Stipules linear, (2–)3–6 mm. long, caducous. Peduncles filiform, (1–)2–4 cm. long, sparsely pilose to tomentulose; pedicels filiform, (3–)5–10 mm. long in flower, 6–10 (–13) mm. long in fruit. Cup pilose to tomentulose. Sepals 1·5–2 mm. long, externally pilose to tomentulose. Petals 1·4–1·8 mm. long, whitish. Disk whitish. Fruit pendulous, globose to slightly obovoid-globose, 5–7 mm. long, pilose and with crowded acute glandular tubercles ± 0·3 mm. long, ripening through shades of red, said to be blue-black when ripe. Fig. 5.

UGANDA. Ankole District: Isingiro, Gayaza, 30 Jan. 1956, *Harker* 180 !; Mbale District: Bugishu, Buginyanya, 29 Aug. 1932, *A. S. Thomas* 365 !; Masaka District: Kabula, Biwolobo, Sept. 1945, *Purseglove* 1810 !

KENYA. Embu District: Chuka, *M.D. Graham* in *A.D.* 1703 !; Machakos District: Kilungu location about 5·5 km. N. of Nunguni, E. side of mountain slopes, 11 June 1967, *Mwangangi* 53 !; Kericho District: Sotik, 15 June 1953, *Verdcourt* 949 !

TANGANYIKA. Ngara District: Bushubi, Nyakisasa, 5 Jan. 1961, *Tanner* 5704 !; Arusha District: Ngurdoto National Park, Senato, 13 Oct. 1965, *Greenway & Kanuri* 12131 !; Mbeya District: Usafwa, 14 Mar. 1914, *Stolz* 2590 !

DISTR. **U**1–4; **K**1, 3–6; **T**1–5, 7; northern Ethiopia, Somali Republic, Rwanda, Burundi, Congo, Zambia, Rhodesia, Mozambique, South Africa (Natal)

HAB. A wide variety of situations, probably commonest in wooded grassland, forest margins and secondary bushland generally, growing over bushes; 100–2300 m.

SYN. *Rhamnus mystacinus* Ait., Hort. Kew. 1: 266 (Aug. 1789)
Mystacinus cirrhiferus Raf., Sylv. Tellur.: 30 (1838), *nom. superfl., illegit.*, based on *R. mystacinus* Ait.
Helinus mystacinus (Ait.) Steud. forma *pilosiusculus* Radlk. in Abh. Naturwiss. Ver. Bremen 8: 387 (1883), as "*pilosiuscula*". Type: Ethiopia, Tigre Province, near Adua [Aduwa], 28 May 1837, *Schimper* I. 155 (M, holo. !)
H. mystacinus (Ait.) Steud. forma *tomentosus* Radlk. in Abh. Naturwiss. Ver. Bremen 8: 387 (1883), as "*tomentosa*". Type: Ethiopia, Tigre, Mt. Sholoda, 28 Oct. 1837, *Schimper* I. 363 (M, holo. !, BR, iso. !)
?H. mystacinus (Ait.) Steud. var. *somalensis* Engl., V.E. 3 (2): 316 (1921). Type: "nördlichen Somalland", locality and collector not specified (B, holo. †)

6. RHAMNUS

L., Sp. Pl.: 193 (1753) & Gen. Pl., ed. 5: 89 (1754)

Trees or shrubs (or scandent shrubs). Leaves alternate (or opposite but not in East Africa), petiolate. Stipules present, small, free, usually soon deciduous. Flowers usually in axillary cymes or these reduced to fascicles or solitary flowers, or rarely in more elaborate bracteate or non-bracteate panicles or thyrses, 5- or 4-merous, usually perfect but in some species dioecious (? not in East Africa), perigynous. Petals present or absent. Disk thin, lining the cup. Ovary 3-celled or reportedly rarely 4-celled (or 2-celled but not in East Africa). Style usually 3-partite about half the length. Fruit a drupe with 3 (or 2 but not in East Africa) or rarely 4 free 1-seeded stones.

A genus of perhaps 150 conservatively circumscribed species of tropical and temperate regions.

R. purshiana DC. is reported to be occasionally cultivated in East Africa.

Petioles 3–10 mm. long; leaf-blades elliptic or elliptic-oblong, 3–10 cm. long, 15–40 mm. wide; sepals 5; petals 5 or often absent . . . . . 1. *R. prinoïdes*
Petioles 1–3(–5) mm. long; blades obovate or narrowly obovate or broadly oblanceolate, rarely elliptic, (1–)2–3(–4·4) cm. long, 7–20 mm. wide; sepals 4; petals 4 . . . . 2. *R. staddo*

1. **R. prinoïdes** *L'Hérit.*, Sert. Angl.: 6, t. 9 (1788); Hemsl. in F.T.A. 1: 382 (1868); P.O.A. C: 255 (1895); Fiori in Rev. Agric. Subtrop. & Tropic. 5, Suppl. 12: 14 (1912); Engl. in Z.A.E.: 489 (1914) & in V.E.: 311 (1921); Staner in B.J.B.B. 15: 403 (1939); T.T.C.L.: 468 (1949); I.T.U., ed. 2: 325 (1952); Brenan in Mem. N.Y. Bot. Gard. 8: 191–256 (1953); Exell & Mendonça, C.F.A. 2: 31 (1956); Verdc. in B.J.B.B. 27: 358 (1957); F.W.T.A., ed. 2, 1: 670 (1958); E.P.A.: 501 (1960); Evrard in F.C.B. 9: 433 (1960); K.T.S.: 391 (1961); F.F.N.R.: 227 (1962); R.B. Drummond in F.Z. 2: 427, fig. 89/B (1966). Type: South Africa, Cape of Good Hope, *Masson* (G-DC, holo. !, M, probable iso. !)

Shrubs or trees up to 8 m. tall, the branches sometimes scrambling or sprawling. Year-old branches blackish, lenticellate, glabrate; youngest twigs usually minutely pubescent with dark spreading hairs. Leaf-blades elliptic-ovate to ovate-elliptic or narrowly so, (1–)2–10 cm. long, (8–)14–40 mm. wide, rounded to cuneate at base, acuminate, firm, lustrous, nearly glabrous except for minute hairs on the major nerves beneath, glandular-serrulate, on each side of midrib with 4–6 secondary nerves which are impressed above, prominulent beneath; petioles (3–)5–10 mm. long. Stipules subulate, 3–5 mm. long, minutely pubescent, caducous. Flowers rarely solitary, usually in fascicles of 2–3(–8), 5-merous; pedicels filiform, minutely pubescent, 3–10 mm. long in flower, 7–15 mm. in fruit. Cup minutely pubescent externally, 2 mm. wide. Sepals ± 2 mm. long. Petals spatulate, ± 1 mm. long, or usually absent in East African populations. Fruit nearly globose, 5–6 mm. thick, reddish. Fig. 6.

UGANDA. Kigezi District: Mt. Mgahinga, 22 Oct. 1929, *Snowden* 1586 !; Mbale District: Butandiga, 14 July 1924, *Snowden* 911 !; Mengo District: 16 km. on Kampala–Bombo trail, Jan. 1938, *Chandler* 2108 !
KENYA. Elgon, 22 Mar. 1931, *Lugard* 576 !; N. Kavirondo District: W. Kakamega Forest Reserve, 13 July 1960, *Paulo* 553 !; Masai District: Olopito near Rotian, 15 Feb. 1952, *Mackay* in *Bally* 8095 !
TANGANYIKA. Masai District: Longido Mt., 16 Jan. 1936, *Greenway* 4376 !; Moshi District: N. Kilimanjaro, Kitenden, 4 Sept. 1950, *Carmichael* 39 !; Lushoto, 14 Mar. 1956, *Makwila* 13 !
DISTR. **U**1–4; **K**1, 3–6; **T**2–7; Ethiopia to South Africa (Cape Province) and W. to Cameroun and Angola
HAB. Forest, evergreen bushland and thicket; 700–3700 m.

SYN. *R. pauciflora* A. Rich., Tent. Fl. Abyss. 1: 137 (1847), as "*pauciflorus*"; Suessenguth in E. & P. Pf., ed. 2, 20d: 65 (1953). Type: Ethiopia, Semien, Adesula [Add'silam], 4 Sept. 1838, *Schimper* II.1276 (P, holo. !, L, M, S, iso. !)
R. prinoïdes L'Hérit. var. *acuminata* Kuntze, Rev. Gen. 3(2): 39 (1898), as "*acuminatus*". Type: South Africa, Transvaal, Pretoria, 17 Feb. 1894, *Kuntze* (NY, holo. !)
R. prinoïdes L'Hérit. var. *obtusifolia* Kuntze, Rev. Gen. 3(2): 39 (1898), as "*obtusifolius*". Type: South Africa, Cape Province, Cathcart, 25 Feb. 1894, *Kuntze* (NY, holo. !)

NOTE. Mr. J. B. Gillett (in litt.) says the species probably occurs in the Teita Hills (**K**7), but there are no specimens to document this.

2. **R. staddo** *A. Rich.*, Tent. Fl. Abyss. 1: 138 (1847); Hemsl. in F.T.A. 1: 382 (1868); Evrard in F.C.B. 9: 434 (1960); K.T.S.: 391, fig. 77 (1961);

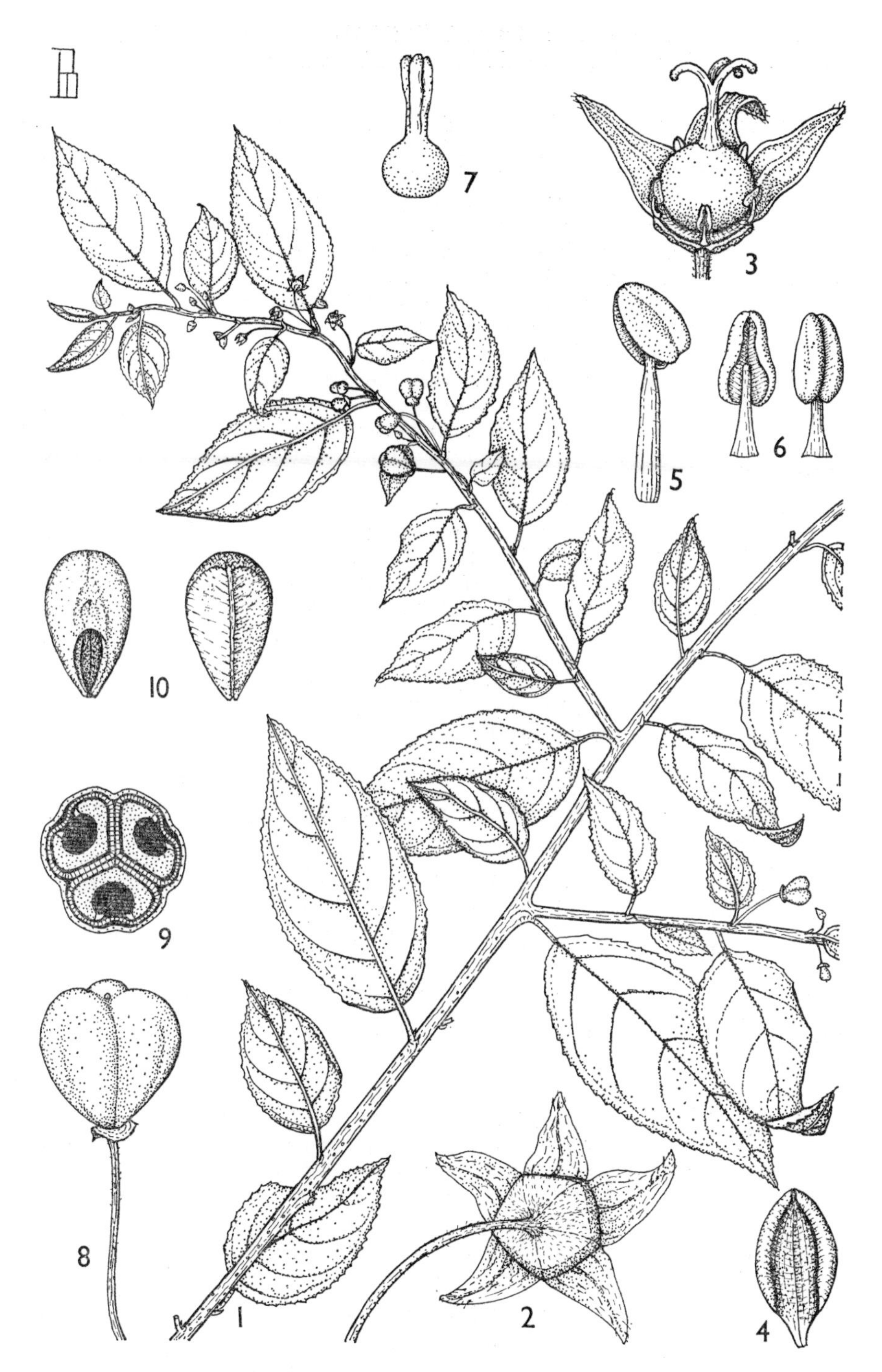

FIG. 6. *RHAMNUS PRINOIDES*—**1,** habit, × ⅔; **2,** flower, viewed from beneath, × 6; **3,** flower, with two sepals removed, × 10; **4,** petal, × 20; **5,** long stamen, × 20; **6,** short stamen, back and front views, × 20; **7,** ovary, × 20; **8,** fruit, × 4; **9,** transverse section of fruit, × 4; **10,** stone, two views, × 4. 1–3, from *Willan* 234; 4, from *Stolz* 451; 5, 7, from *Glover et al.* 1602; 6,8–10, from *Trapnell* 60/458.

R.B. Drummond in F.Z. 2: 428, t. 89/C (1966). Type: Ethiopia, Eritrea, Wojerat [Ouodgerate], *Quartin Dillon & Petit* (P, holo. !)

Deciduous shrubs or small trees, 0·5–7 m. tall, sometimes scandent. Bark smooth, grey; slash cream; year-old branches dark brown-grey, slender, glabrate; youngest twigs densely to less commonly sparsely hispidulous with grey hairs 0·1–0·2 mm. long; short shoots (many-noded twigs) often produced late in the growing season; elongated twigs occasionally thorn-tipped. Leaf-blades obovate or narrowly obovate or broadly oblanceolate, rarely elliptic, (1–)2–3(–4·4) cm. long, (5–)6–15(–20) mm. wide, cuneate, blunt to acute and mucronate, dark olive green or brown-green (at least when dry), glabrous or hispidulous on the major nerves beneath and occasionally above or rarely hispidulous on both surfaces with grey hairs 0·1–0·2 mm. long, obscurely glandular-serrulate or occasionally with the teeth reduced to vanishingly small size, with 5 to 8 pairs of obscure secondary nerves; petioles (1–)2–5(–8) mm. long, usually at first hispidulous and later glabrescent. Stipules filiform to subulate, 1–3 mm. long, caducous to subpersistent. Flowers borne singly at the nodes, often appearing at the nodes of the short-shoots of the previous season and thus in somewhat fascicle-like aggregations, greenish or sometimes said to be cream colour or orange, fragrant; pedicels usually glabrous, 1–2 (–4) mm. long in flower, in fruit ± 2(–4·5) mm. long. Cup glabrous or rarely hispidulous. Sepals 4, deltoid, 1–1·7 mm. long, glabrous or rarely hispidulous on back. Petals 4, linear to oblanceolate, ± 0·5–0·8 mm. long, not clasping, caducous. Stamens 4, said to be brownish when fresh. Fruit nearly globose, ± 5–5·5 mm. long, reddish or scarlet when ripe, with 3 stones or by abortion fewer, each stone on drying eventually capable of colubrinoid dehiscence, revealing an orange or brick red inner lining. Seed ± 4–5 mm. long, pale brownish, smooth, with a ventral groove.

UGANDA. Ankole District: Lutobo Hill, Sept. 1941, *Eggeling* 4530 !; Kigezi District: Kamwezi, May 1950, *Eggeling* 5875 !; Masaka District: Kabula, 14 Mar. 1936, *Michelmore* 1334 !

KENYA. Northern Frontier Province: Moyale, 28 Apr. 1952, *Gillett* 12968 !; Naivasha District: Ol Kalou, 16 Dec. 1948, *Bally* 6518 !; Kisumu-Londiani District: shoulder of Limutet [Limutit], 26 Sept. 1953, *Drummond & Hemsley* 4466 !

TANGANYIKA. Mbulu District: Mt. Hanang, NE. slope, 8 Feb. 1946, *Greenway* 7670 !; Arusha District: Mt. Meru, E. side, 4 Oct. 1932, *B.D. Burtt* 4141 !; Njombe, 3 Jan. 1932, *Lynes* D.g. 171b !

DISTR. **U**2, 4; **K**1, 3–6; **T**1–3, 7; Congo, Rwanda, Somali Republic, Ethiopia (including Eritrea), Rhodesia; reportedly in Yemen

HAB. Forest margins and upland evergreen bushland; 1000–3000 m. often at high altitudes

SYN. *R. infusionum* Del. in Ferr. & Galin., Voy. Abyss. 3: 111 (1847), *nom. superfl.*, *illegit.*, based on *R. staddo* A. Rich.

R. holstii Engl., P.O.A. C: 255 (1895) & in Z.A.E.: 489 (1914) & V.E. 3(2): 312 (1921); T.T.C.L.: 468 (1949); Verdc. in B.J.B.B. 27: 357 (1957). Type: Tanganyika, Usambara Mts., Kwambugu, *Holst* 3789 (B, holo.†, M, iso. fragm. !)

R. uhligii Engl. in E.J. 40: 552 (1908) & V.E. 3(2): 312 (1921); T.T.C.L.: 468 (1949); Verdc. in B.J.B.B. 27: 357 (1957); K.T.S.: 391 (1961) Type: Tanganyika, Arusha District, Olkokola Mts., *Uhlig* 415 (B, holo. †)

R. deflersii Fiori, Rev. Agr. Subtrop. & Trop. 5, Suppl. 12: 15 (1912). Type: Ethiopia, Eritrea, near Acrur [Acrour], *Schweinfurth & Riva* 1040 (B, holo. †, P, S, iso. !)

R. staddo A. Rich. var. *deflersii* (Fiori) Engl., V.E. 3 (2): 312 (1921)

R. staddo A. Rich. var. *espinosus* Engl., V.E. 3 (2): 312 (1921), *nom. nud.*

R. rhodesicus Suesseng. in Mitt. Bot. Staatssamml. München 1 (6): 181 (1953). Type: Rhodesia, Rusape, *Dehn* R40/52 (M, holo. !, BR, K, iso. !)

R. sp. sensu Verdc. in B.J.B.B. 27: 357 (1957)

7. SCUTIA

Brongn., Mém. Fam. Rhamn.: 55 (1826), *nom. conserv.*

Glabrous or nearly glabrous thorny shrubs or small trees, the thorns either straight (New World) or short and recurved (Africa, etc.). Branchlets and leaves opposite or nearly so; petioles short. Flowers 5-merous, in condensed axillary cymes or axillary fascicles or solitary in the axils. Disk lining the cup, rather thin. Ovary 2–3-celled (very rarely 3-merous in Africa); style very short and slightly lobed. Drupe nearly globose, at maturity pulpy and with 2 or 3 seeds each enclosed in the free endocarpous stone which has no evident ventral pore or slit and is never regularly dehiscent.

A genus of 4 species, 3 in South America plus the present one.

S. myrtina (*Burm. f.*) *Kurz* in Journ. Asiat. Soc. Beng. 44 (2): 168 (1875); T.T.C.L.: 469 (1949); Perr., Fl. Madag. & Com., fam. 123: 4 (1950); I.T.U., ed. 2: 326 (1952); Verdc. in B.J.B.B. 27: 358 (1957); Evrard in F.C.B. 9: 431, fig. 11 (1960); K.T.S.: 391 (1961); F.F.N.R.: 227 (1962); R.B. Drummond in F.Z. 2: 428, t. 89/A (1966). Type: India, Coromandel, collector not stated (G, not found)

Shrubs or rarely small trees 2–5(–10) m. tall, usually scandent. Bole to 3 dm. thick; older trunks with thick dark corky longitudinally cracked bark; younger herbage and young inflorescences puberulent; branchlets angular, usually emerging at right-angles, numerous. Thorns prickle-like, recurveds axillary, solitary (but often 2 per node), 2–10(–15) mm. long. Leaf-blade, ovate to obovate, 2–6 cm. long, 15–40 mm. wide, at base rounded to cuneate, at apex (rarely slightly acuminate to) rounded (sometimes apiculate) or retuse or emarginate, always mucronulate, entire or in the distal two-thirds with a few (up to 9 on each side) indistinct crenulations or appressed teeth, on each side of midrib with 5–8 obscure secondary nerves; petioles 3–10 mm. long. Stipules 2–3 mm. long, quickly deciduous. Cymes often fascicle-like, 2–20-flowered (never more than 1 flower per cyme maturing fruit); peduncles 2–7 mm. long; pedicels 1–2(–3) mm. long in flower, 2–3 mm. in fruit. Sepals 1·2–2 mm. long. Petals 0·7–1 mm. long. Drupe 7–9 mm. long and thick, pallid when very young, ripening through red to purplish black; endocarps ± 6 mm. long, readily separating into 2(–3) free indehiscent stones. Fig. 7.

Uganda. Karamoja District: Mt. Kadam [Debasien], Zebiel, 9 Jan. 1937, *A. S. Thomas* 2228!; Ankole District: Kashari County, stock farm, 1 Feb. 1956, *Harker* 207!; Mengo District: Old Entebbe, Nov. 1932, *Eggeling* 698!

Kenya. Northern Frontier Province: Ngare Ndare, 4 June 1944, *Davidson* in *Bally* 4440!; Masai District: Chyulu Range, North and Centre, Apr. 1938, *Bally* 7675!; Kilifi District: Mariakani, 17 Sept. 1961, *Polhill & Paulo* 497!

Tanganyika. Bukoba District: Minziro Forest, Jan. 1959, *Procter* 1120!; Lushoto District: Shagayu [Shagai] Forest Reserve, Aug. 1955, *Semsei* 2230!; Njombe, Oct. 1931, *Staples* 162!

Zanzibar. Zanzibar I., near Chwaka, 27 Dec. 1933, *Vaughan* 2184!

Distr. **U**1–4; **K**1–7; **T**1–7; **Z**; Congo, Burundi, Zambia, Rhodesia, Malawi, Mozambique, South Africa, Madagascar, Seychelles, Mascarene Is., Ceylon, India, Burma, Thailand, North Vietnam

Hab. In a wide variety of situations from forest margins to bushland, thicket and wooded grassland; from sea-level to 2700 m.

Syn. *Rhamnus myrtina* Burm. f., Fl. Ind.: 60 (1768), as " *myrtinus* "

R. circumscissa L. f., Suppl.: 152 (1781), as " *circumscissus* ". Type: E. Peninsular India, *Koenig* (LINN, holo.!, C, iso.!)

R. capensis Thunb., Prodr. 1: 44 (1794); Radlk. in Abh. Bot. Ver. Brandenb. 8: 389 (1883). Type: South Africa, Cape of Good Hope, *Thunberg* 5438 (UPS, holo.!)

R. lucida Roxb., Fl. Indica, ed. 1, 2: 353 (1824). Type: Mauritius, *Roxburgh* (BM, holo.!)

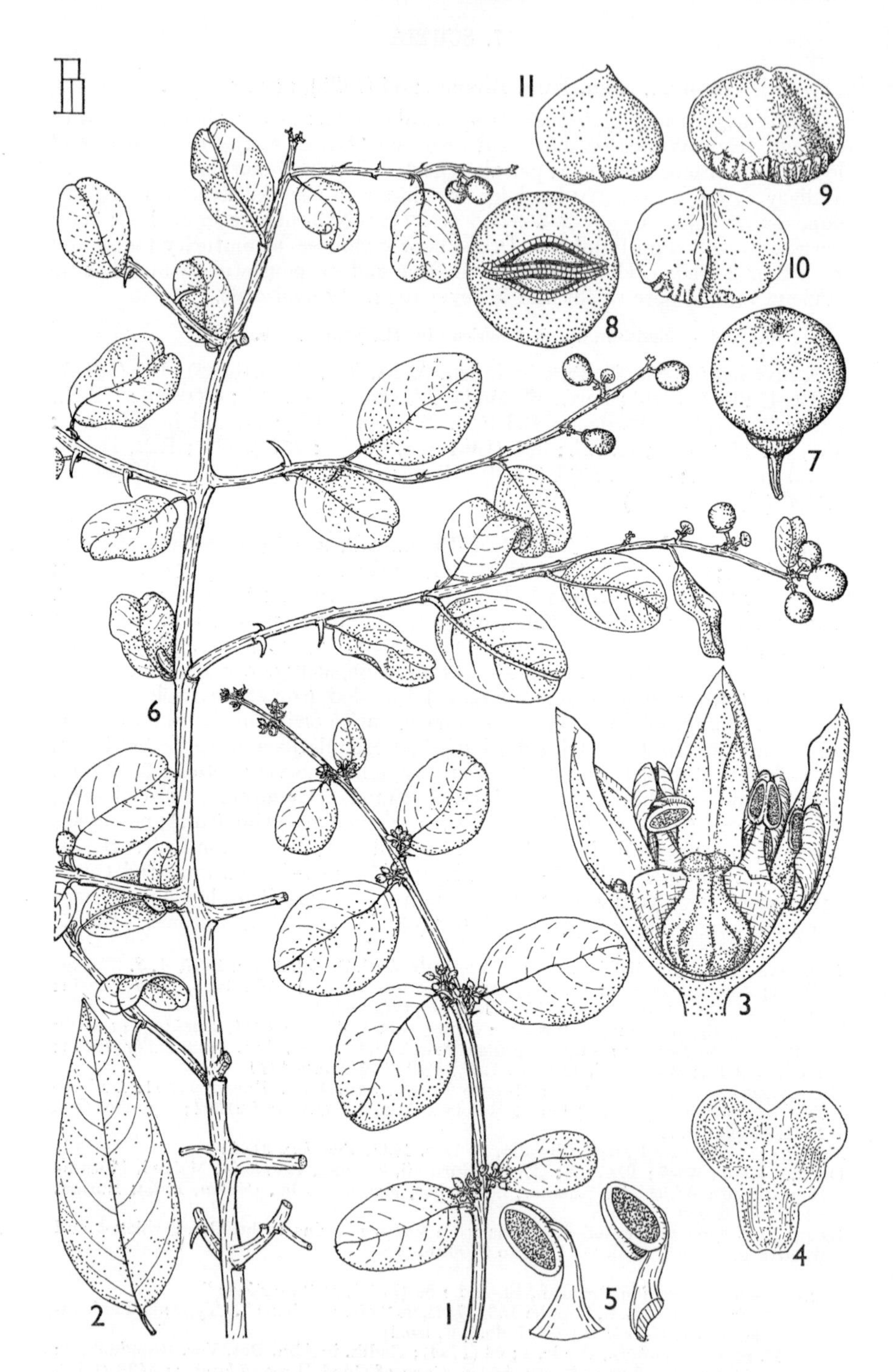

FIG. 7. *SCUTIA MYRTINA*—**1,** flowering branch, × $\frac{2}{3}$; **2,** leaf, × $\frac{2}{3}$; **3,** vertical section of flower, × 10; **4,** petal, spread out, × 20; **5,** stamens, × 20; **6,** fruiting branch, × $\frac{2}{3}$; **7,** fruit, × 2; **8,** transverse section of fruit, × 3; **9,** stone, viewed from outer side, × 3; **10,** same, from inner side, × 3; **11,** seed, viewed from outer side, × 3. 1, 6, from *Kokwaro* 241; 2–5, from *Glover et al.* 996; 7–11, from *Glover et al.* 2043.

Scutia commersonii Brongn., Mém. Fam. Rhamn.: 56. t. 4 (1826). Type: Réunion [Bourbon], *Commerson* (P, holo. !)
S. indica Brongn., Mém. Fam. Rhamn.: 56 (1826), *nom. superfl.*, *illegit.*, based on *Rhamnus circumscissa* L. f.
S. capensis (Thunb.) G. Don, Gen. Syst. 2: 33 (1832)
S. lucida (Roxb.) G. Don, Gen. Syst. 2: 33 (1832)
S. capensis (Thunb.) G. Don var. *parvifolia* Eckl. & Zeyh., Enum. Pl. Afr. Austr. Extratrop.: 130 (1835). Type: South Africa, Port Elizabeth and Algoa Bay, *Ecklon* (FI, holo. !, C, FR, K, L, M, P, W, iso. !)
S. natalensis Krauss in Flora 27: 346 (1844). Type: South Africa, Durban [Natal Bay], *Krauss* 322 (TUB, holo. !, BM, K, iso. !)
S. obcordata Tul. in Ann. Sci. Nat., sér. 4, 8: 116 (1857). Type: Madagascar, Ste. Marie I., 1849, *Boivin* 2644 (P, holo. !)
S. indica Brongn. var. *oblongifolia* Engl. in E.J. 19, Beibl. 47: 37 (1894). Type: Tanganyika, Kilimanjaro, Marangu, *Volkens* 1395 (B, holo. †, BM, BR, E, K, iso. !)
S. indica Brongn. var. *grandispina* Engl., V.E. 3(2): 210 (1921). Type: Kenya, Mau Escarpment, collector not stated (B, holo. †)
S. buxifolia Hutch. & Moss in K.B. 1928: 272 (1928), *non* Reiss. (1861), *nom. illegit.* Type: Kenya, Laikipia, R. Ngobit, *Gardner* in *F.D.* 1477 (K, holo. !, EA, iso. !)
S. hutchinsonii Suesseng. in Lilloa 4: 135 (1939). Type: as for *S. buxifolia* Hutch. & Moss
S. myrtina (Burm. f.) Kurz var. *oblongifolia* (Engl.) Evrard in F.C.B. 9: 432, fig. 11/D–F (1960)

NOTE. This species is somewhat variable but the variation is chaotic and nearly co-extensive throughout the vast range, so that the recognition of infraspecific taxa is not practicable.

8. ZIZIPHUS

P. Mill., Abbrev. Gard. Dict. [unpaginated] (1754)

Trees or shrubs (or scandent shrubs or almost lianes but not in East Africa), often armed by spinous stipules (or by branch-thorns but not in East Africa), the branchlets often zig-zag. Leaves alternate or opposite, petiolate; blades often 3- or 5-nerved from the base and penninerved above, often ovate or elliptic, with serrulate margins (or entire margins but not in East Africa). Stipules present, often at some nodes transformed into short spines, some of which are recurved, the stipules small, subulate and quickly deciduous at those nodes where they are not so transformed. Flowers 5-merous, bisexual, usually in axillary cymes or thyrselets, perigynous. Petals present (or absent but not in tropical Africa), clasping the stamens or at least at some stages clasping the filaments. Disk usually thickened near, but never coherent to, the sides of the ovary. Ovary 2–4-celled (3 cells are very rare and 4 cells absent in East Africa). Fruit a drupe with a single stone; cells 2 or 3 (or 4 but not in East Africa).

A genus of about 86 species of the temperate and tropical parts of the world, most common in dry areas.

A number of species of *Ziziphus* yield edible fruit, and some have been cultivated and selected since ancient times. A specimen from a plant cultivated at Magrotto, E. Usambara Mts., Tanganyika (*Omari Chambo* 8) has been tentatively determined as *Z. incurva* Roxb., a species indigenous to India, Burma and Nepal. It may on the other hand represent the related *Z. angustifolia* (Miq.) van Steenis of SE. Asia and Malesia. A specimen from an abandoned coastal garden in M'toni, Zanzibar (*Faulkner* 3110) has been determined, apparently correctly, as *Z. rugosa* Lam., a species indigenous to India.

Leaf-blades, or at least most of the blades on any one specimen, averaging longer than 15 mm.; petioles averaging 3 mm. or usually more in length:
Branchlets unarmed; leaf-blades usually with a minute dense tomentum beneath; each mar-

ginal tooth accompanied by a minute bundle of hairs of a character distinct from that of the other leaf-pubescence . . . . . . 1. *Z. pubescens*
Branchlets armed or rarely unarmed; leaf-blades tomentose to glabrescent beneath; pubescence in the vicinity of the leaf-teeth of the same nature as that elsewhere on the blade:
Leaves tomentose beneath at maturity:
Leaf-blades markedly asymmetrical at base (fig. 8/4):
Leaf-blades with nerves not markedly impressed above; pubescence of the dull upper surface persistent . . . 2. *Z. mucronata* subsp. *rhodesica**
Leaf-blades with nerves markedly impressed above; pubescence (if any) of the often slightly lustrous upper surface almost entirely deciduous eventually . . 3. *Z. abyssinica*
Leaf-blades nearly symmetrical at base (fig. 8/1) . . . . 4. *Z. mauritiana*
Leaves glabrous or merely finely pubescent beneath at maturity:
Leaf-blades elliptic, never acuminate (fig. 8/5); flowers tomentulose . . . . . 5. *Z. spina-christi*
Leaf-blades ovate, tending to be acuminate (fig. 8/6); flowers nearly glabrous . . 2. *Z. mucronata* subsp. *mucronata*
Leaf-blades 8–15(–20) mm. long, 5–7 mm. wide; petioles 1–2(–3) mm. long . . . . . 6. *Z. hamur*

1. **Z. pubescens** *Oliv.* in Trans. Linn. Soc., ser. 2, 2: 330 (1887); T.T.C.L.: 470 (1949); I.T.U., ed. 2: 329 (1952); Verdc. in B.J.B.B. 27: 355 (1957); Evrard in F.C.B. 9: 442, t. 45 (1960); K.T.S.: 395 (1961); R.B. Drummond in Bol. Soc. Brot., sér. 2, 39: 59 (1965) & in F.Z. 2: 424 (1966). Type: Tanganyika, Kilimanjaro, *H.H. Johnston* (K, holo.!)

Unarmed shrubs (rarely described by collectors as clambering) or trees to 10(–20) m. tall. Bole short; slash-wood salmon-red to whitish; sapwood whitish; bark fairly rough to shallowly reticulately fissured, dark grey-brown. Branches spreading and often drooping; branchlets lenticellate, grey-brown, the youngest with a sericeous tomentum or ascending to spreading yellow-brown hairs 0·3–0·4 mm. long. Leaf-blades narrowly ovate or less commonly narrowly oblong-obovate, (2–)3–5(–8) cm. long, (1·4–)2–2·5(–3) cm. wide, at base slightly and roundly unequal-sided, shortly acuminate to less commonly merely acute or rarely blunt, finely and antrorsely serrate (the glandular point of each tooth accompanied by a bundle of greyish hairs 0·1–0·2 mm. long), beneath with a sericeous tomentum of spreading yellow-brown hairs ± 0·3 mm. long (especially dense along the nerves), above puberulent to glabrous, 3-nerved basally; petioles 3–6 mm. long, tomentose. Stipules subulate, 1–2 mm. long, caducous. Flowers yellow-green, numerous in axillary cymes; cymes 5–10 mm. long and broad, 8–20 flowered (only 1 flower per cyme producing fruit, rarely 2); peduncles 1–4 mm. long; pedicels 1–2 mm. long in flower, 3–6 mm. long in fruit. Sepals ± 1·5 mm. long, spreading in flower. Petals ± 1 mm. long, very narrow, spreading in flower. Ovary 2-celled; stigmas 2. Fruit nearly globose (prolate when immature), yellow or red-yellow, 7–10 mm. long. Fig. 8/3, p. 25.

* See note under subsp. *mucronata*, p. 26.

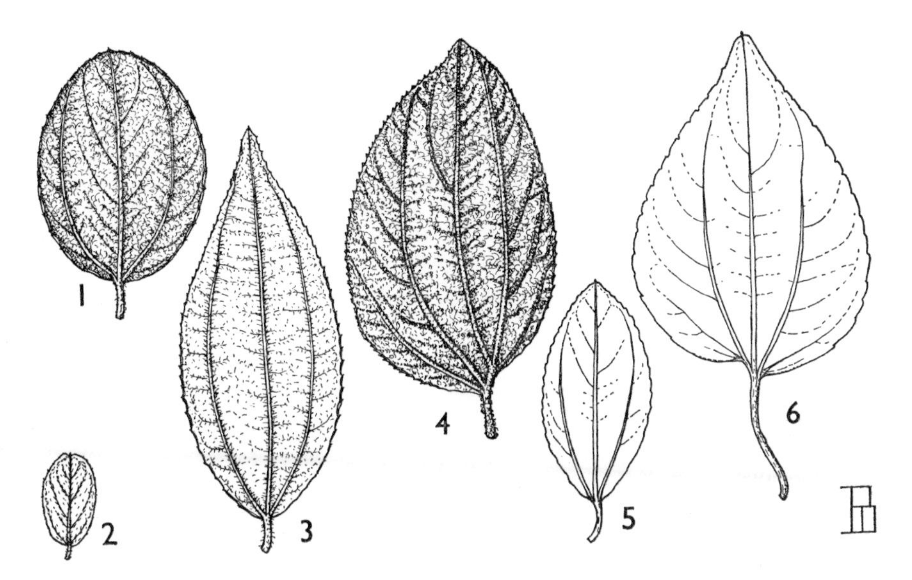

FIG. 8. Leaves of species of *Ziziphus*, × 2/3. **1,** *Z. mauritiana*, from *R. M. Graham* 245; **2,** *Z. hamur*, from *Gillett* 13314; **3,** *Z. pubescens*, from *Dale* K786; **4,** *Z. abyssinica*, from *Hazel* 641; **5,** *Z. spina-christi*, from *Forest Dept.* 448; **6,** *Z. mucronata* subsp. *mucronata*, from *Leippert* 5247.

UGANDA. Lango District: Kyoga, Namasale, 20 May 1948, *For. Dept.*!; Bunyoro District: near Nile, Jan. 1947, *Dale* 528!; Mengo District: Bugerere, 5·6 km. N. of Bale, 29 June 1956, *Langdale-Brown* 2125!

KENYA. Kitui District: 32 km. S. of Kitui, 10 June 1955, *Bogdan* 4030!; Lake Victoria, Wangama I., 11 Jan. 1955, *Argyle* 103!; Masai District: Tsavo National Park West, Mzima Springs, 24 June 1953, *Bally* 8976!

TANGANYIKA. Kwimba District: Mantare, Nov.-Dec. 1933, *Gillman* 4!; Masai/Arusha District: Monduli, Mto-wa-Mbu, 21 Nov. 1954, *Matalu* 3250!; Kilwa District: W. of Lungonya R., 11 Nov. 1967, *Rodgers* 88!

DISTR. **U**1–4; **K**4–7; **T**1–3, 6; Ethiopia, Congo, Sudan, Zambia, Rhodesia, Malawi, Mozambique; probably also Cameroun and Angola

HAB. Usually riparian in wooded grassland; 0–1000 m.

SYN. ?*Z. espinosa* Buettn. in Verh. Bot. Ver. Brandenb. 32: 48 (1890). Type: Congo Leopoldville, Coquela, near Underhill, Tondoa, *Buettner* 493 (B, holo. †)

Z. jujuba (L.) Gaertn. var. *scheffleri* Engl., V.E. 3(2): 305 (1921). Type: Kenya, Machakos District, Kibwezi, *Scheffler* 391[1] (B, holo. †, BM, E, K, L, P, S, Z, iso.!)

Celtis polyclada Peter, F.D.O.-A. 2, Anhang: 1, t. 18/1 (1932). Type: Tanganyika, Lushoto District, Handëi, Malamba [Maramba]–Lugongo, *Peter* 21020 (B, lecto.!, K, photo.!)

NOTE. Verdcourt in B.J.B.B. 27: 356 (1957) indicates the occurrence in Northern Frontier Province (**K**1) but no specimen has been seen from there or from the Turkana Province.

2. **Z. mucronata** *Willd.*, Enum. Hort. Berol.: 251 (1809); Hemsl. in F.T.A. 1: 380 (1868); P.O.A. C: 255 (1895); Hutch. & Bruce in K.B. (1941); T.T. C.L.: 470 (1949); I.T.U., ed. 2: 328 (1952); Exell & Mendonça, C.F.A. 2: 28 (1956); Verdc. in B.J.B.B. 27: 355 (1957); F.W.T.A., ed. 2, 1: 669 (1958); Evrard in F.C.B. 9: 441 (1960); K.T.S.: 394 (1961); R.B. Drummond in Bull. Soc. Brot., sér. 2, 39: 57 (1965) & in F.Z. 2: 422 (1966). Type: South Africa, inland from Cape of Good Hope, *Lichtenstein* (B-WILLD, No. 4663, holo.!)

Shrubs or small trees to 15(–30) m. tall, armed with spinous stipules or rarely unarmed. Bark of bole dark grey, smooth to rough to corrugated;

slash-wood crimson, soft. Year-old branches often zig-zag. Leaf-blades ovate, (2–)3–6(–8) cm. long, (1·3–)2–3·5(–4·7) cm. wide, at base often markedly asymmetrical, rounded or rarely very shallowly cordate, acute, often acuminate or rarely obtuse, serrulate, 3-nerved from base, the nerves of the dull upper surface only slightly if at all impressed; petioles 2–7 mm. long. Cymes (5–)10–15 mm. long, and about as thick, (3–)7–25-flowered; peduncles 1–3 mm. long; pedicels 1–3 mm. long in flower, 3–5 mm. long in fruit. Sepals 1·5–2 mm. long, spreading in flower. Petals 1–1·5 mm. long, spreading in flower. Ovary 2-celled; stigmas 2. Fruit globose, 12–20 mm. thick, reddish or reddish brown at maturity.

subsp. **mucronata**; R.B. Drummond in Bol. Soc. Brot., sér. 2, 39: 57 (1965) & in F.Z. 2: 422 (1966)

Plants nearly glabrous or with sparse pubescence on the nerves of the leaf-blades.

UGANDA. Karamoja District: Amudat, Oct. 1944, *Dale* 411!; Ankole District: Ruizi R., 12 Nov. 1950, *Jarrett* 465!
KENYA. Northern Frontier Province: W. Marsabit Mt., June 1958, *T. Adamson* 21!; Masai District: Mosiro [Musilo], 19 Aug. 1960, *Paulo* 689!; Teita District: Voi R., 31 Jan. 1953, *Bally* 8641!
TANGANYIKA. Tanga District: Kivindani, 3 Feb. 1966, *Faulkner* 3736!; Mpwapwa District: Kongwa, 28 Feb. 1949, *Anderson* 367!; Iringa, 18 Mar. 1932, *Lynes* I.g. 224!
DISTR. **U**1, 2; **K**1–7; **T**1–7; Senegal to Arabia and S. to South Africa, in the drier parts, also Madagascar
HAB. Open woodland or wooded grassland; sea-level to 2000 m.

SYN. *Z. bubalina* Schult. in Roem. & Schult., Syst. 5: 334 (1819), *nom. superfl., illegit.*, based on *Z. mucronata* Willd.
Z. baclei DC., Prodr. 2: 20 (1825). Type: Senegal, 1820, *Bacle* (G-DC, holo.!)
Z. adelensis Del., Rochet d'Héricourt, Sec. Voy. Mer Rouge: 341 (1846). Type: Ethiopia, Shoa [Choa], Fare [Farré]–Alio Amba [Aleyon amba], 1845, *Rochet d'Héricourt* 20 (P, holo.!)
Z. mitis A. Rich., Tent. Fl. Abyss. 1: 137 (1847). Type: Ethiopia, Shoa [Choa], *Quartin Dillon & Petit* (P, holo.!, K, W, iso.!)
Z. mucronata Willd. var. *pubescens* Sond. in Fl. Cap. 1: 476 (1860). Types: South Africa, Durban [Port Natal], *Gueinzius* & *T. Williamson* (? TCD, isosyn.)
Z. mucronata Willd. var. *glabrata* Sond. in Fl. Cap. 1: 476 (1860). Types: South Africa, Transvaal, Magaliesberg, *Burke & Zeyher* 311 & 313 (? TCD, isosyn.)
Z. mucronata Willd. var. *glabrata* Sond. forma *subcordata* Kuntze, Rev. Gen. 3 (2): 39 (1898). Type: South Africa, Cape Province, Modderrivier Station, 10 Feb. 1894, *Kuntze* (NY, holo.!, K, iso.!)
Z. mucronata Willd. var. *glabrata* Sond. forma *obliqua* Kuntze, Rev. Gen. 3 (2): 39 (1898). Type: South Africa, Natal, Durban, Berea Hills, Mar. 1894, *Kuntze* (NY, holo.!, K, Z, iso.!)
Z. mucronata Willd. var. *glauca* Schinz, Vier. Nat. Gesellsch. Zürich 29: 195 (1903). Type: South West Africa, Keetmanshoop, *Schinz* 837 (Z, lecto.!, K, iso.!)
Z. mucronata Willd. var. *inermis* Engl., V.E. 3 (2): 307 (1921); T.T.C.L.: 470 (1949). Type: Tanganyika, Rungwe District, Kyimbila, *Stolz* 1816 (B, holo.!, B, L, Z, iso.!)
Z. mucronata Willd. forma *pubescens* (Sond.) Chiov., Fl. Somala 1: 127 (1929)
Z. madecassus Perr. in Not. Syst. Paris 11: 17 (1943). Type: SW. Madagascar, R. Onilahy near Maroamalona, *Perrier* 19235 (P, holo.!)

NOTE. Subsp. *rhodesica* R.B. Drummond is reported from S. Tanganyika in F.Z. 2: 423 (1966), but no specimen has been seen. It differs from subsp. *mucronata* by the persistent coarse pale brown tomentum on the leaf-blades (denser beneath), and occurs in Zambia, Congo (Katanga), Angola, Botswana and South West Africa.

Some plants of this subspecies so closely resemble *Z. abyssinica* as to suggest the origin through hybridization or genetic contamination of *Z. mucronata* subsp. *mucronata* by *Z. abyssinica* with much subsequent backcrossing to *Z. mucronata* subsp. *mucronata*. However, the persistence of the upper pubescence possibly is not explained by this suggestion. An alternative hypothesis is that subsp. *rhodesica* shows closer relationship to *Z. abyssinica* than to *Z. mucronata*. The persistence of upper pubescence and the small size and bluntness of the leaf-blades can then be considered a manifestation of simple neoteny, since these characters are also seen in young branches of *Z. abyssinica*. Any taxonomic reshuffling must await a thorough analysis in the field.

3. **Z. abyssinica** *A. Rich.*, Tent. Fl. Abyss. 1: 136 (1847); T.T.C.L.: 469 (1949); I.T.U., ed. 2: 326, fig. 69/a, c–e (1952); Brenan in Mem. N.Y. Bot. Gard. 8: 239 (1953); Exell & Mendonça, C.F.A. 2: 29 (1954), pro parte, excl. syn. Engl. & Gilg; Verdc. in B.J.B.B. 27: 353 (1957); F.W.T.A., ed. 2, 1: 669 (1958); Evrard in F.C.B. 9: 440 (1960); E.P.A.: 498 (1960); K.T.S.: 393 (1961); F.F.N.R.: 228 (1962); R.B. Drummond in F.Z. 2: 420 (1966). Type: Ethiopia, Tigre, Djeladjeranne, 20 Jan. 1840, *Schimper* III.1694 (P, holo. !, BR, FI, FR, GOET, K, L, M, S, UPS, W, Z, iso. !)

Shrubs or small trees (2–)3–6(–13) m. tall, armed; trunk to 25 cm. thick; bark grey, very deeply furrowed or fissured; slash-wood deep red; sapwood reddish brown. Branchlets pubescent, zig-zag. Leaf-blades ovate to broadly ovate, (3–)5–8(–13) cm. long, (2–)3–5(–9) cm. broad, at base rounded and very asymmetric, obtuse to acute or even acuminate, 3-nerved from the base, above when young glabrous or finely pubescent, eventually glabrous and with somewhat impressed venation, beneath often tomentose to nearly glabrous and with prominulent veins; petioles (3–)5–8(–12) mm. long, tomentose. Cymes 1–2 cm. long and broad, (6–)10–25-flowered; peduncles 2–6(–10) mm. long, tomentose; pedicels 0·5–3 mm. long in flower, 3–6(–9) mm. long in fruit, tomentose. Sepals 1·5–2 mm. long, dorsally tomentose. Petals 1–1·5 mm. long. Ovary 2-celled. Drupe 2–3 cm. thick, reddish when mature. Stone 2-seeded. Fig. 8/4, p. 25, & 9, p. 28.

Uganda. W. Nile District: Moyo, Dec. 1931, *Brasnett* in *F.D.* 305 !; Teso District: Serere, May 1932, *Chandler* 668 !; Mengo District: Entebbe, *Fyffe* 128 !
Kenya. Northern Frontier Province: Moyale, 9 Oct. 1952, *Gillett* 14018 !; Machakos District: Kibwezi–Ithaba plains, 15 May 1938, *Bally* 7674 !; Teita District: Mwatate, *H.H. Johnston* !
Tanganyika. Moshi District: E. Kilimanjaro, Himo R., 24 Jan. 1936, *Greenway* 4486 !; Mbeya District: 1·6 km. S. of Songwe river on Mbeya–Tunduma road, 5 Jan. 1963, *Boaler* 806 !; Masasi District: Ndanda, 25 May 1943, *Gillman* 1442 !
Distr. **U**1, 3, 4; **K**1–5, 7; **T**1, 2, 4, 5, 7, 8; Senegal to Ethiopia and S. to Angola, Rhodesia and Mozambique
Hab. Scattered-tree grassland; 400–2200 m.

Syn. [*Z. jujuba* sensu Hemsl. in F.T.A. 1: 379 (1868), pro parte, quoad syn. *Z. abyssinica*; Eyles in Trans. Roy. Soc. S. Afr. 5: 406 (1916), pro parte, quoad specim. *Rogers*; Exell in J.B. 65, Suppl. Polypet.: 80 (1927), *non* (L.) Gaertn.]
Z. jujuba (L.) Gaertn. forma *obliquifolia* Engl., Hochgebirgsfl. Trop. Afr. (in Abh. Königl. Preuss. Akad. Wiss. 1891): 294 (1892). Type: Ethiopia, Tigre, Djeladjeranne, 20 Jan. 1840, *Schimper* III.1694 (B, syn. †, BR, FI, FR, GOET, K, L, M, P, S, UPS, W, Z, isosyn. !)
Z. jujuba (L.) Gaertn. var. *obliquifolia* Engl. in E.J. 30: 351 (1902). Type: Tanganyika, Rungwe District, mouth of R. Luferio, *Goetze* 873 (B, holo.†, BR, L, iso. !)
Z. jujuba (L.) Gaertn. var. *nemoralis* Sim, For. Fl. Port. E. Afr.: 35 (1909), ex char. Type: Mozambique, *Sim* (not located)
Z. mauritiana Lam. var. *abyssinica* (A. Rich.) Fiori, Bosch. e piant. legn. Eritr.: 234 (1912); A. Chev. in Rev. Intern. Bot. Appliq. et d'Agric. Trop. 27: 477 (1947)
Z. baguirmiae A. Chev. in Étud. Fl. Afr. Centr. Franç. 1: 59 (1913), *nom. subnud.* Type: Chad, Baguirmi, between Abou Gher and Arahil, *A. Chevalier* 9685 bis (P, holo. !, BR, L, W, iso. !)
[*Z. mauritiana* sensu Staner in B.J.B.B. 15: 398 (1939): Wild, Guide Fl. Vict. Falls: 151 (1952); Suesseng., E. & P. Pf., ed. 2, 20d: 124 (1953), pro parte, *non* Lam.]
Z. atacorensis A. Chev. in Rev. Intern. Bot. Appliq. et d'Agric. Trop. 27: 479 (1947). Type: Dahomey, Atacora Mts., foot of Tanguiéta cliff, *A. Chevalier* 24090 in part (P, holo. !, K, iso. !)
Z. atacorensis A. Chev. var. *oblongifolia* A. Chev. in Rev. Intern. Bot. Appliq. et d'Agric. Trop. 27: 480 (1947). Type: Dahomey, upper Volta, near Konkobiri, *A. Chevalier* 24367 in part (P, holo. !, BR, K, iso. !)

Note. Exell and Mendonça, C.F.A. 2: 29 (1956), include in synonymy here the *Z. jujuba* (L.) Gaertn. var. *aequilaterifolia* Engl. & Gilg (in Warb., Kunene-Sambesi-Exped.: 292 (1903); type: Angola, Huila, between Goudkopje and Kakele, *Baum* 190 (B,

FIG. 9. *ZIZIPHUS ABYSSINICA*—**1,** branch with flowers and fruits, × 1; **2,** leaf, × 1; **3,** flower bud, × 6; **4,** flower, × 6; **5,** same, showing atypical tetramerous form, × 4; **6,** mature flower with petals fallen, × 6; **7,** petal, side view, × 12; **8,** stamen, × 12; **9,** portion of infructescence, × 1. All from *Fyffe* 128.

holo. †, BM, E, K, M, Z, iso. !)); this collection seems to be determinable as *Z. zeyherana* Sond., a species reaching its northern limit in southern Angola and Rhodesia

A specimen from Ufipa District, Sumbawanga–Mkunde[Nkunde], 29 Nov. 1949 *Bullock* 1954, has all the earmarks of *Z. abyssinica*, but is said to represent a shrub to 6 dm. tall forming miniature thickets in grassland. The habit is that of *Z. zeyherana* Sond. This indicates the need for careful field observation on the relationship of that species to *Z. abyssinica*.

Occasionally the pubescence of the upper surface of the leaves at least partially persists. The persistent pubescence is usually found in specimens with blunt leaf-blades less than 1·5 times as long as broad, e.g. Kenya, Machakos/Kitui District, Yatta, 56 km. E. of Thika, 11 June 1961, *J.G. Williams* in *E.A.H.* 12331; Tanganyika, Ufipa District, Muse–Sumbawanga road, 10 Nov. 1963, *Richards* 18383, Mbeya District, Igawa, Jan. 1962, *Procter* 1957, and Kondoa District, Bereku Ridge, 13 Jan. 1928, *B.D. Burtt* 1060. The two apparently linked characters of persistent pubescence and leaf shape signal similarity to *Z. mucronata* subsp. *rhodesica* R.B. Drummond, which, however, has not yet been found in genetically pure state in East Africa, so far as is known.

4. **Z. mauritiana** *Lam.*, Encycl. 3: 319 (1789); U.O.P.Z.: 492, fig. (1949); T.T.C.L.: 469 (1949); I.T.U., ed. 2: 328, fig. 69/b (1952); Verdc. in B.J.B.B. 27: 354 (1957); F.W.T.A., ed. 2, 1: 668 (1958); Evrard in F.C.B. 9: 440 (1960); E.P.A.: 498 (1960); K.T.S.: 394 (1961); F.F.N.R.: 228 (1962); R.B. Drummond in F.Z. 2: 420 (1966). Type: Mauritius [Ile de France], *Sonnerat* (P-LA, holo. !)

Shrubs or small trees 3–8(–16) m. tall, armed with spinous stipules or rarely unarmed; bark greyish. Branchlets densely but minutely pubescent zig-zag. Leaf-blades elliptic to ovate to nearly orbicular, 3–8 cm. long, 1·5–5 cm. broad, at base rounded and symmetrical or nearly so, obtuse, beneath densely tomentose; petioles 5–10 mm. long. Cymes 1–2 cm. long and broad, few–many-flowered; peduncles 1–4 mm. long, tomentose; pedicels 2–4 mm. long in flower, 3–6 mm. in fruit, tomentose. Sepals 1·5–2 mm. long, dorsally tomentulose. Petals 1–1·5 mm. long. Ovary-cells and seeds 2. Drupe globose to ellipsoidal, 1–2 cm. thick. Fig. 8/1, p. 25.

UGANDA. Acholi District: Agoro, Apr. 1943, *Purseglove* 1501 !; Lango District: Rom, *Eggeling* 2362 !; Karamoja District: Napak, 27 May 1940, *A. S. Thomas* 3564 !

KENYA. Northern Frontier Province: Sololo, 3 Aug. 1952, *Gillett* 13678 !; Turkana District: Lokitaung, 22 May 1953, *Padwa* 201 !; Kwale District: Mwachi, *R.M. Graham* 245 !

TANGANYIKA. Tanga District: Amboni Estate, 26 Dec. 1959, *G.R. Williams* 727 !; Pangani District: N. of Pangani, 20 Jan. 1937, *Greenway* 4870 !; Lindi, 30 Apr. 1903, *Busse* 2339 !

ZANZIBAR. Zanzibar I., Mangapwani, 23 Jan. 1929, *Greenway* 1126 !; SW. of Pemba, Panza I., 13 Feb. 1929, *Greenway* 1404 !

DISTR. **U**1, 3; **K**1–3, 6, 7; **T**3, 6, 8; **Z**; **P**; almost universally cultivated and escaped in tropical parts of the world

HAB. In cultivation and other disturbed areas near settlements and along roads; 0–1400 m.

SYN. *Rhamnus jujuba* L., Sp. Pl.: 194 (1753). Type: Ceylon, *Hermann* 89 (BM, holo. !)

Ziziphus jujuba (L.) Gaertn., Fruct. 1: 203 (1788); Lam., Encycl. 3: 319 (1789); Hemsl. in F.T.A. 1: 379 (1868), pro parte; Boiss., Fl. Orient. 2: 13 (1872); Sim, For. Fl. Port. E. Afr.: 35 (1909), pro parte, excl. var. *nemoralis*; Engl., Hochgebirgsfl. Trop. Afr. (in Abh. Königl. Preuss. Akad. Wiss. Berl. 1891): 294 (1892) & in Ann. Ist. Bot. Roma 7: 19 (1910), *non* Mill. (1768), *nom. illegit.*

Rhamnus mauritiana Soyer-Willemet in Uster's Neue Ann. Bot. 18: 20 (1796). Type: Mauritius, The Pouce, St. Pierre plains, etc., collector not stated (specimen not located)

Ziziphus tomentosa Poir. in Lam., Encycl., Suppl. 3: 192 (1813). Type: Santo Domingo I., *Poiteau* (P, holo. !)

Z. rotundata DC., Prodr. 2: 21 (1825). Type: Mauritius, collector not stated (G-DC, holo. !)

Z. orthocantha DC., Prodr. 2: 21 (1825). Type: Senegal, 1820, *Bacle* (G-DC, holo. !)

Z. aucheri Boiss., Diagn., sér. 1, 1 (2): 5 (1843). Type: Iran, Bushir [Bouchir], *Aucher* 4320 (G, holo. !, FI, K, P, iso. !)

Z. jujuba (L.) Gaertn. var. *stenocarpa* Kuntze, Rev. Gen. 1: 121 (1891). Type: India, Delhi, cultivated, 15 Dec. 1875, *O. Kuntze* (NY, holo.!)
Z. jujuba (L.) Gaertn. forma *aequilaterifolia* Engl., Hochgebirgsfl. Trop. Afr. (in Abh. Königl. Preuss. Akad. Wiss. 1891): 294 (1892), *nom. nud.*, *non* sensu Engl. & Gilg in Warb., Kunene-Sambesi-Exped.: 292 (1903), descr.
Z. mauritiana Lam. var. *orthocantha* (DC.) A. Chev. in Rev. Intern. Bot. Appliq. et d'Agric. Trop. 27: 477 (1947)

NOTE. This widely cultivated jujube is not known to be indigenous to Africa. It is probably originally from the Middle East or the Indian subcontinent and is represented widely in the tropics and subtropics by several varieties with rather large fruits and almost tree stature. Some shrubbier, smaller-fruited, smaller-leaved plants found away from cultivation may represent the wild forms from which the cultivated plants arose in ancient times. Such plants have been called *Z. jujuba* (L.) Gaertn. var. *fruticosa* Haines, For. Fl. Chota Nagpur: 270 (1910) and *Z. mauritiana* Lam. var. *deserticola* A. Chev., Rev. Intern. Bot. Appliq. et d'Agric. Trop. 27: 477 (1947). The taxonomy of cultivated jujubes is difficult. Selection, atavism and broadscale human spreading of various genetic strains are probably involved. Possibly also hybridization is involved. Keay, F.W.T.A., ed. 2, 1: 669 (1958), cites specimens from Dahomey and Nigeria which appear to him to be hybrids between *Z. mauritiana* and *Z. spina-christi* (L.) Desf. Probable hybrids between these two species occur in India and Pakistan and are the basis of the *Z. jujuba* (L.) Gaertn. var. *hysudrica* Edgew. in J.L.S. 6: 201 (1862) or *Z. hysudrica* (Edgew.) Hole in Indian Forester 55: 505 (1918).

5. **Z. spina-christi** (*L.*) *Desf.*, Fl. Atl. 1: 201 (1798); Del. in Cailliaud, Voy. Meroé 4: 378 (1827); Hemsl. in F.T.A. 1: 380 (1868); V.E. 3 (2): 307 (1921); Perr. in Not. Syst. Paris 11: 27 (1943) & Fl. Madag. & Comor., fam. 123: 12 (1950); U.O.P.Z.: 492 (1949); T.T.C.L.: 470 (1949); Verdc. in B.J.B.B. 27: 355 (1957); E.P.A.: 499 (1960). Type: Israel, Jerusalem, *Hasselquist* (LINN, 262.38, holo.!)

Trees to 10 m. tall, armed with spinous stipules or rarely unarmed; bark greyish. Youngest branchlets minutely but rather densely pubescent. Leaf-blades narrowly ovate to rarely elliptic, 2–6 cm. long, 1·2–3·2 cm. wide, at base rounded and nearly symmetrical, obtuse, minutely but densely pubescent beneath when young, tending to almost completely glabrescent at maturity; petioles (3–)10–15(–20) mm. long. Cymes ± 1 cm. long and thick, (5–)10–25-flowered; peduncles 1–2 mm. long; pedicels 1–3 mm. long in flower, 3–6 mm. long in fruit, at first minutely woolly. Sepals minutely woolly dorsally, ± 2 mm. long. Petals ± 1·5 mm. long. Drupe globose, ± 1 cm. thick; cells and seeds 2, rarely 3. Fig. 8/5, p. 25.

var. **spina-christi**

Leaf-blades averaging 4–6 cm. long. Petioles averaging much longer than 3 mm.

UGANDA. Karamoja District: below Moroto, 6 Mar. 1936, *Michelmore* 1259!
TANGANYIKA. Shinyanga, May 1950, *Forest Dept.* 448!; Kilwa District: Kilwa, Singoni (?=Singino), 31 May 1906, *Braun* in *Herb. Amani* 1299! & Kilwa Kisiwani, 10 Apr. 1909, *Kraenzlin* in *Herb. Amani* 3000!
ZANZIBAR. Pemba I., Kukuu, 17 Dec. 1930, *Greenway* 2748!
DISTR. **U**1; **T**1, 3, 6, 8; **Z**; **P**; widespread in Near and Middle East and drier parts of Africa; cultivated and an escape in East Africa
HAB. Near settlements in disturbed areas, occasionally roadsides; sea-level to 1300 m.

SYN. *Rhamnus spina-christi* L., Sp. Pl.: 195 (1753)
Ziziphus africana P. Mill., Gard. Dict. ed. 8 [unpaginated] (1768). Type: not specified
Rhamnus nabeca Forsk., Fl. Aegypt.-Arab.: 204 (1775), pro parte, quoad var. b. Types: Yemen, *Forsskål*, one flowering and one fruiting specimen collected in different seasons (C-Forsskål, holo.!)
Ziziphus nabeca (Forsk.) Lam., Encycl. 3: 320 (1789), as "*napeca*"
Z. spina-christi (L.) Desf. var. *inermis* DC., Prodr. 2: 20 (1825). Type: Lebanon [Levant], *Olivier* 1822 (G-DC, holo.!)
Z. sphaerocarpa Tul. in Ann. Sci. Nat., sér. 4, 8: 119 (1857). Type: Comoro Is., Mayotte, Pamanzi, Longoni Bay, *Boivin* 3364 (P, holo.!)

Z. iroënsis A. Chev. in Étud. Fl. Afr. Centr. Franç. 1: 59 (1913). Type: Chad, Moyen Chari, Lake Iro, Gouré [Kouré], *A. Chevalier* 9040 (P, holo.!, K, iso.!)
Z. spina-christi (L.) Desf. var. *longipes* Engl., V.E. 3 (2): 307 (1921). Type: Tanganyika, Uzaramo District, Kwale I., *Busse* 3144 (B, holo. †, EA, iso.!)
Z. spina-christi (L.) Desf. var. *mitissima* Chiov., Fl. Somala 2: 139 (1932); E.P.A.: 499 (1960). Type: Somali Republic (S.), Afgoi, *Scassellati* 172 (FI, holo.!)

Note. The typical variety is not known to be indigenous to Africa except perhaps in the extreme north-eastern part. Even there it may have been introduced as a cultivated plant in ancient times.

var. **microphylla** *A. Rich.*, Tent. Fl. Abyss. 1: 136 (1847). Ethiopia, Tigre, Djeladjerane, 19 Apr. 1841, *Schimper* III.1798 (P, holo.!, B, BM, E, FI, GOET, K, L, M, iso.!)

Leaf-blades or most of them on each specimen only 2–3 cm. long. Petioles averaging 3 or 4 mm. long.

Kenya. Turkana District: between Lokitaung and Lodwar, 2 Aug. 1938, *Pole-Evans & Erens* 1593!; Masai District: Magadi, 1959, *Maini*! & 25 Feb. 1963, *Glover & Cooper* 3505!
Distr. **K**2, 6; Ethiopia, Somali Republic; probably widespread elsewhere in N. Africa; Yemen
Hab. Apparently rare along semi-desert washes; 600–1000 m.

Syn. *Rhamnus nabeca* Forsk., Fl. Aegypt.-Arab.: 204 (1775), pro parte, quoad var. a. Type: Yemen, Môr, Jan. 1763, *Forsskål* (C-Forsskål, holo.!)
?*Ziziphus amphibia* A. Chev., Rev. Bot. Appliq. 27: 480 (1947). Type: Mali, San, R. Bani, *A. Chevalier* 1084 (P, holo.!, K, iso.!)

Note. In contrast to the nominate variety, the var. *microphylla* appears to be indigenous to Africa. Its relationship to *Z. hamur* Engl. is in need of investigation.

6. **Z. hamur** *Engl.* in Ann. Ist. Bot. Roma 7: 19 (1910) & V.E. 3(2): 307 (1921); Hutch. & Bruce in K.B. 1941: 129 (1941); Verdc. in B.J.B.B. 27: 353 (1957); E.P.A.: 498 (1960); K.T.S.: 394 (1961). Type: Somali Republic (S.), Uebi R., *Robecchi-Bricchetti* 186 (B, holo.†, FI, Z, iso.!)

Shrubs 1–2(–3) m. tall, armed; bark apparently dark greyish. Branchlets zig-zag, often arcuate-recurved, puberulent. Leaves often fascicled at the nodes; blades oblong, 8–15(–20) mm. long, 5–7(–13) mm. wide, shortly but densely pilosulous all over, at base rounded to broadly cuneate, acute to rounded, minutely serrulate to nearly entire, grey-green; petioles 1–3 mm. long. Cymes much reduced, fascicle-like or glomeruliform, ± 5 mm. long and thick, 2–4-flowered; peduncles essentially absent; pedicels 2–3·5 mm. long in flower, 6–7 mm. in fruit, pilosulous. Sepals pilosulous dorsally, 1–1·2 mm. long. Petals ± 1 mm. long, white. Drupe globose, 7–10 mm. thick, black at maturity. Fig. 8/2, p. 25.

Kenya. Northern Frontier Province: Mandera, 25 May 1952, *Gillett* 13314!
Distr. **K**1; Ethiopia, Somali Republic
Hab. In semi-desert scrub on shallow red sandy soil over limestone; ± 350 m.

Note. This species has close affinity on the one hand to the Saharan *Z. lotus* (L.) Desf. subsp. *saharae* Maire in Bull. Soc. Hist. Nat. Afr. Nord 20: 179 (1929), and on the other to the Arabian *Z. leucodermis* (Bak.) O. Schwartz in Mitt. Inst. Bot. Hamb. 10: 152 (1939). That the three taxa are distinct from one another, at least at the species level, is unlikely. A thorough field study is desirable.

9. BERCHEMIA

DC., Prodr. 2: 22 (1825), *nom. conserv. propos.*

Phyllogeiton (Weberb.) Herzog in Bot. Centralbl., Beih. 15: 168 (1903)

Unarmed trees or shrubs (or scandent shrubs or lianes or twining vines but not in Africa). Leaves opposite or subopposite (or alternate but not in Africa), petiolate; blades entire, penninerved, the secondary nerves often describing numerous close parallel arcs. Stipules present, intra-axillary (or free but not in Africa). Cymes often reduced to axillary fascicles (or in most species, but not the African ones, aggregated into raceme-like thyrses and the thyrses in turn aggregated into pyramidal panicles). Flowers bisexual, 5-merous. Ovary superior though in some species closely invested by the disk, 2-celled; style short and usually quickly deciduous after anthesis. Drupe elongate, usually 2–3 times as long as thick, with a single elongate 2-celled, usually 2-seeded, stone.

A genus of 22 species, one in North America, two in Africa and the rest in southern Asia.

B. discolor (*Klotzsch*) *Hemsl.* in F.T.A. 1: 381 (1868); Sim, For. Fl. Port. E. Afr.: 35 (1909); Bak. f. in J.L.S. 40: 45 (1911); Eyles in Trans. Roy. Soc. S. Afr. 5: 407 (1916); Burtt Davy, Fl. Pl. & Ferns Transv. 2: 470 (1932); T.T.C.L.: 466 (1949); Codd, Trees and Shrubs Kruger Nat. Park: 111 (1951); O. B. Mill. in Journ. S. Afr. Bot. 18: 50 (1952); I.T.U., ed. 2: 323 (1952); F.F.N.R.: 227 (1962); R.B. Drummond in F.Z. 2: 425, t. 88/A (1966). Type: Mozambique, Manica e Sofala, Sena, *Peters* (B, holo.†, K, P, iso. !)

Shrubs or trees usually no more than 10 m. tall but reported occasionally to 25 m. and with trunks 5–7 dm. thick; wood very hard and heavy; bark deeply checked and tending to shed in sheets. Younger branches conspicuously lenticellate; branchlets 2–20 cm. long, leafy, glabrous to densely pubescent with short spreading whitish hairs. Leaves opposite or nearly so; earliest (lowest) blades of emerging branchlets smallish, elliptic-obovate and blunt, later ones larger, elliptic to ovate-oblong, acute (2–)3–5(–9) cm. long, (1·5–)2–3·5(–6) cm. wide, at the base cuneate or usually rounded, at the apex blunt to acute, green above and glabrous or minutely pubescent near the midrib, beneath slightly paler and microvesiculate and glabrous to densely pubescent with short whitish hairs, on each side of midrib with 6–8 (–9) secondary nerves; petioles (4–)8–10(–13) mm. long, glabrous to pubescent. Stipules intra-axillary, 2–4 mm. long, subulate, united about half the length, all but the basal 0·5–1 mm. quickly deciduous. Flowers solitary or usually in fascicles of 2–6 in the axils; pedicels 3–5(–7) mm. long in flower, 4–7(–10) mm. in fruit. Sepals 2–3 mm. long, spreading to reflexed at anthesis. Petals 1·8–2·2 mm. long, spreading or weakly arcuate-ascending. Style 1 mm. long, bifid a fourth to a third the length. Disk free from the ovary though thick near its base. Fruit 12–20 mm. long, 7–11 mm. thick, yellow. Fig. 10.

UGANDA. Karamoja District: Akoret, June 1955, *Philip* 725 !
KENYA. Northern Frontier Province: Dandu, 17 Mar. 1952, *Gillett* 12566 !; Baringo District: 17 km. on Marigat–Kabarnet road, 30 Oct. 1964, *Leippert* 5253 !; Teita District: Voi, Mar. 1937, *Dale* in *F.D.* 3676 !
TANGANYIKA. N. Tabora, 24 Jan. 1938, *Lindeman* 559 !; Mpwapwa, 15 Dec. 1935, *Hornby* 736 !; Morogoro R., 7 Jan. 1951, *Wigg* 936 !
DISTR. **U**1; **K**1–4, 6, 7; **T**1–6, 8; Yemen, Ethiopia and Somali Republic south to Angola, South West Africa and South Africa, also Madagascar
HAB. Widespread in thicket, semi-desert grassland and wooded grassland (less abundant but reaching its largest size in stream-valleys or riverine forests); 0–2000 m.

SYN. *Scutia discolor* Klotzsch in Peters, Reise Mossamb. Bot. 1: 110, t. 21 (1861)
Phyllogeiton discolor (Klotzsch) Herzog in Bot. Centralbl. Beih. 15: 169 (1903); Suesseng. in E. & P. Pf., ed. 2, 20d: 140 (1953); Exell & Mendonça, C.F.A. 2: 30 (1954); Coates Palgrave, Trees Centr. Afr.: 367 (1956); Verdc. in B.J.B.B. 27: 356 (1957); K.T.S.: 390 (1961)
Adolia discolor (Klotzsch) Kuntze, Rev. Gen. 1: 117 (1891)
Araliorhamnus punctulata H. Perr. in Not. Syst. Paris 11: 15 (1943). Type: W. Madagascar, between Soalala and Lake Kinkony (Ambongo), *Perrier* 6035 (P, holo. !)
A. vaginata H. Perr. in Not. Syst. Paris 11: 16 (1943). Type: W. Madagascar, Maronfandilia Forest near Morondava, *Perrier* 6010 (P, holo. !)

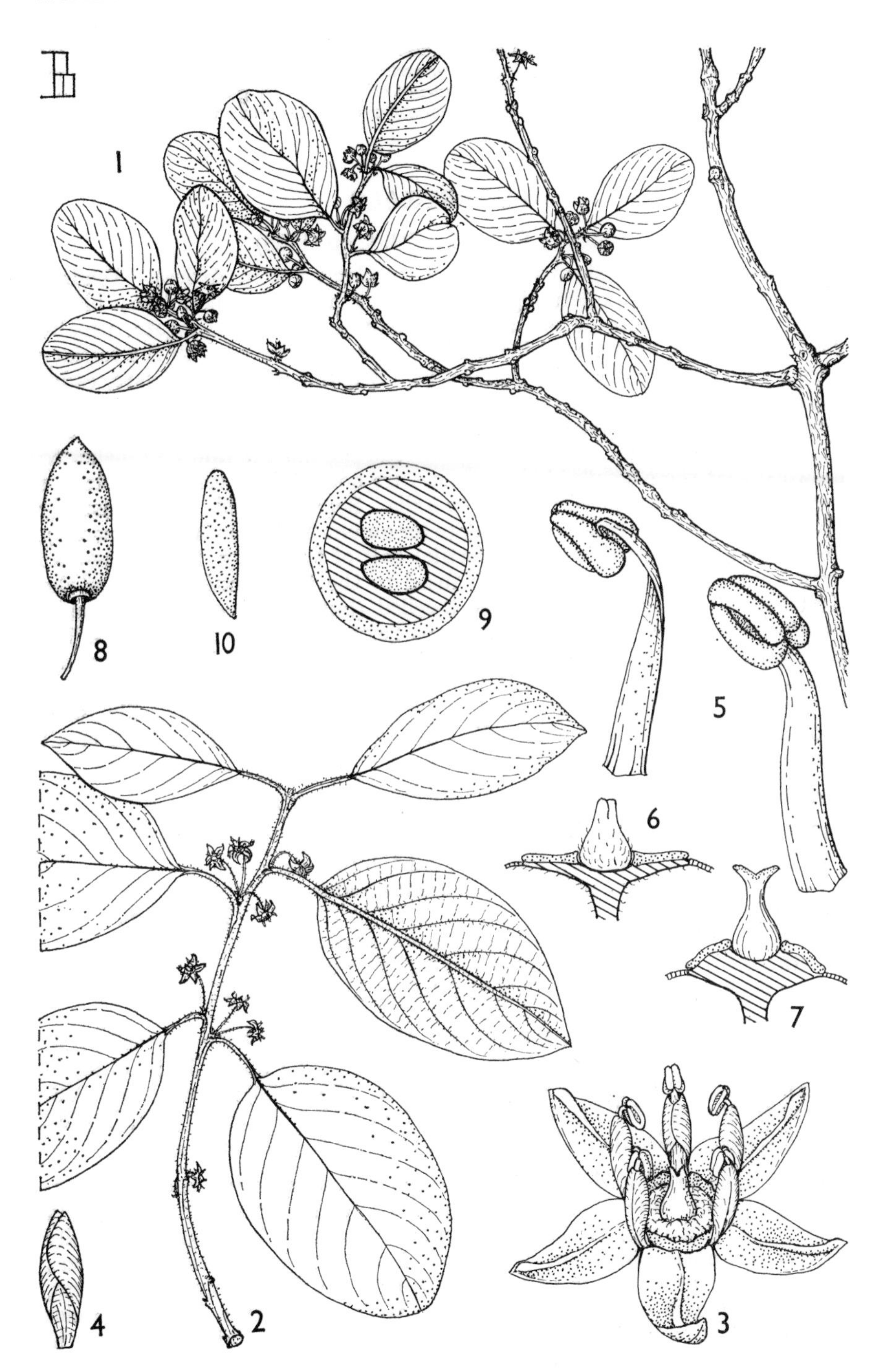

FIG. 10. *BERCHEMIA DISCOLOR*—**1, 2,** flowering branches, showing variation, × $\frac{2}{3}$; **3,** flower, × 8; **4,** petal, front view, × 10; **5,** stamens, × 20; **6, 7,** gynoecium of short- and long-styled flowers respectively, with disk and receptacle cut away to show ovary, × 10; **8,** fruit, × 1; **9,** transverse section of fruit, × 3; **10,** seed, × 2. 1, 3–5, from *Eggeling* 2961; 2, 7, from *Wigg* 936; 6, from *Gillett* 12566; 8–10, from *Hemming* 325.

VARIATION. The variability in pubescence is remarkable but I can discern no other correlated characters and thus do not propose a formal designation for the pubescent plants. A few scattered plants (e.g. *Swynnerton* H.59/36 and *Lindemann* 795) show the acute leaves characteristic of many (but not all) of the Madagascan populations, but again there seems to be no correlation with other variables and no formal recognition is deemed necessary.

10. VENTILAGO

Gaertn., Fruct. 1: 223, t. 49 (1788)

Lianes or climbing shrubs (or small trees but not in Africa). Tendrils absent. Branchlets often zig-zag. Leaves alternate, petiolate; blades serrulate or crenulate or nearly entire, elliptic or more elongate, usually rounded at base, acute, acuminate or obtuse, penninerved with the tertiary nerves often rather elegantly percurrent. Stipules minute, subulate, caducous. Flowers 5-merous, perfect, half-epigynous, usually in very densely crowded divaricate axillary cymes (these so condensed as to appear like nodal glomerules), or sometimes the leaves so reduced and the upper branches so crowded that these smaller inflorescences are borne in thyrses or panicles, or very rarely the flowers solitary in the axils. Sepals deltoid. Petals usually clawed and the body concave, cucullate (or absent but not in African species). Disk thick, adnate to the lower (fertile) half of the ovary, later somewhat accrescent and along with the cup fused to the basal sixth to half of the fruit-body (not the wing). Ovary 2-celled (but only 1 cell fertile), half-inferior, the superior moiety (or perhaps better called the style) expanding to hundreds of times its anthetic size in the form of an elongate terminal strap-like wing the plane of which bisects the 2 original ovary cells and which at the tip carries the 2 minute persistent styles. Fruit dry, samara-like, with a basal 1-celled thinly double-walled nearly spherical body, a single small seed and an elongate terminal wing.

A genus of about 30 or 40 species of the Old World tropics.

Inflorescence comprising small nodal glomerules on usually leafy branches, the axes and the flowers rarely minutely puberulent, usually almost perfectly glabrous . . . . 1. *V. africana*
Inflorescence branched and leafless, its branches and the floral cups and backs of sepals densely pubescent with yellowish hairs about 0·1 mm. long . 2. *V. diffusa*

1. **V. africana** *Exell* in J.B. 65, Suppl. Polypet.: 80 (1927); Staner in B.J.B.B. 15: 396 (1939); Exell & Mendonça, C.F.A. 2: 32 (1956); Verdc. in B.J.B.B. 27: 353 (1957); F.W.T.A., ed. 2, 1: 670 (1958); Evrard in F.C.B. 9: 446 (1960); N. Hallé in Fl. Gabon 4: 56 (1962). Type: Angola, Cuanza Norte, Cazengo, Granja de S. Luis, *Gossweiler* 5721 (BM, holo. !)

Stout liane to 30 m. long. Branches somewhat zig-zag; branchlets essentially glabrous, often divaricate. Leaf-blades ovate-lanceolate or narrowly ovate, 5–12 cm. long, 17–55 mm. broad, at base rounded, acute or usually shortly acuminate (with an acumen ± 1 cm. long), essentially glabrous, beneath with prominulent veins, with ± 6 or 7 pairs of secondary nerves, appressed serrate or appressed crenate; petioles 2–5 mm. long. Stipules subulate, 1–2 mm. long, subpersistent, glabrescent. Flowers borne in axillary glomerules of 4–10 flowers at nodes where the leaves are obsolete or often merely reduced to lanceolate bracts 2–4(–7) cm. long, the compound inflorescences 10–15 cm. long; branches vaguely panicle-like, both terminal and lateral on the branch system, when young minutely pubescent in lines;

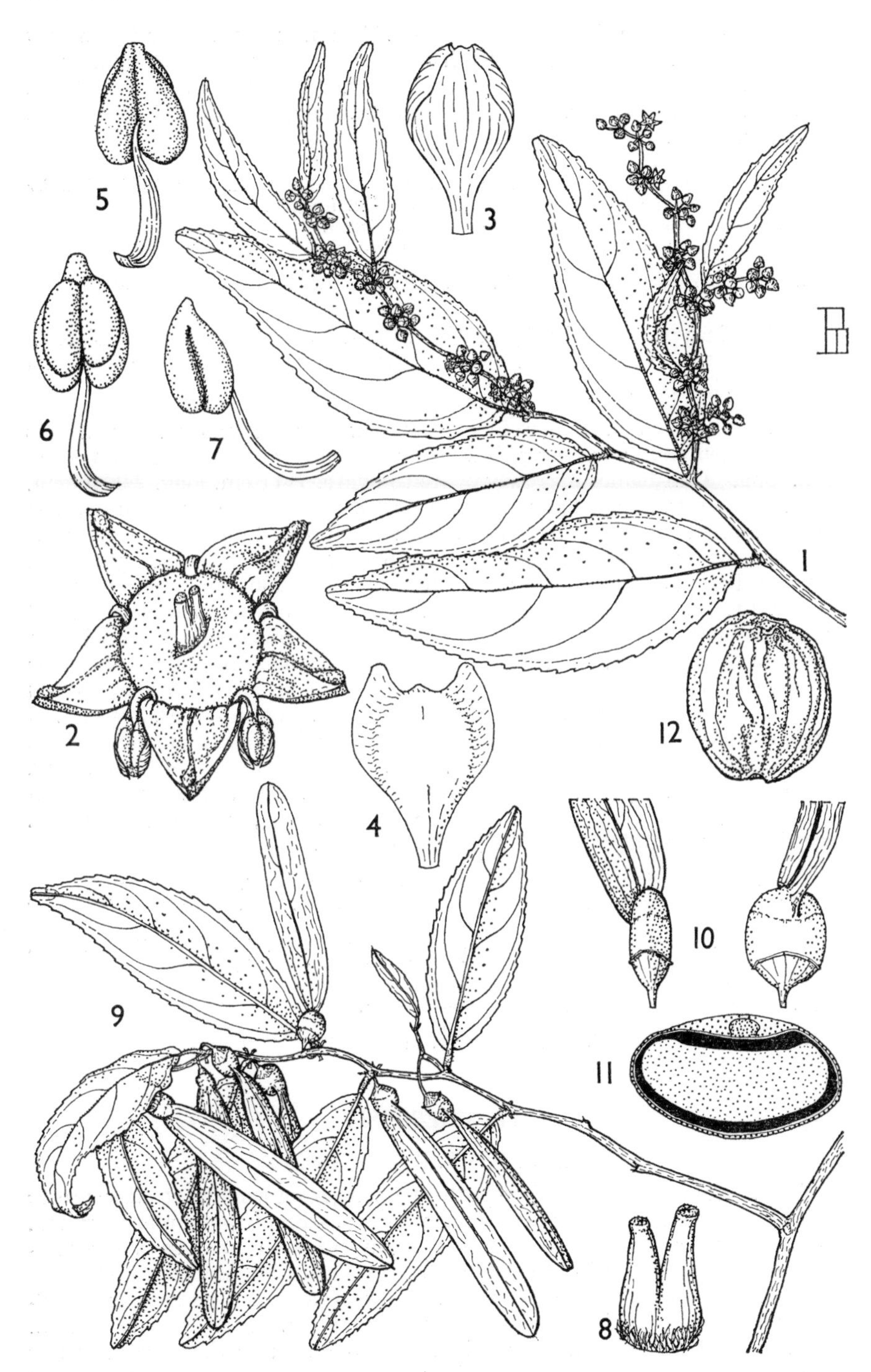

FIG. 11. *VENTILAGO AFRICANA*—**1**, flowering branch, × $\frac{2}{3}$; **2**, flower, × 8; **3**, petal, from inner side, × 20; **4**, same, spread out, from outer side, × 20; **5, 6, 7**, stamen, back, front and side views respectively, × 20; **8**, gynoecium, × 20; **9**, fruiting branch, × $\frac{2}{3}$; **10**, fruit, with wing partly cut away, two views, × 2; **11**, transverse section of fruit, × 6; **12**, seed, × 4. 1–8, from *Chandler* 2025; 9–12, from *Chandler* 1872.

pedicels 1–2(–3) mm. long in flower, 3–4(–5) mm. in fruit, at first minutely pubescent with tawny appressed hairs. Sepals deltoid, 1–1·3 mm. long, glabrous. Petals oblanceolate, ± 1 mm. long, strongly reflexed at anthesis, glabrous. Disk glabrous. Fruit (3–)4–6 cm. long, the wing (7–)8–10 mm. broad, glabrous. Fig. 11, p. 35.

UGANDA. Mengo District: 21 km. on [Kampala–]Entebbe road, Nov. 1937, *Chandler* 2025! & 13 km. on Masaka road, Aug. 1937, *Chandler* 1872!
DISTR. **U4**; Portuguese Guinea to Uganda and Angola
HAB. Apparently in rain-forest at about 1170 m.

SYN. [*V. madraspatana* sensu Engl., V.E. 3(2): 310 (1921), pro parte, *non* Gaertn.]

2. **V. diffusa** (*G. Don*) *Exell*, Cat. Vasc. Pl. S. Tomé: 139 (1944); F.W.T.A., ed. 2, 1: 670 (1958); Evrard in F.C.B. 9: 447 (1960). Type: S. Tomé, *G. Don* (BM, holo.!, K, iso.!)

Stout liane to 35 m. or more in length; bark grey and smooth. Branchlets at least when young densely pubescent with appressed brownish hairs ± 0·1 mm. long. Leaf-blades narrowly ovate to ovate, 7–11 cm. long, 2·9–4·9 cm. wide, at base rounded, apex acute or broadly acuminate (with an acumen ± 1 cm. long), glabrous except for persistent appressed pubescence along the prominulent veins beneath, with about 6 or 7 pairs of secondary nerves, appressed crenulate or appressed serrulate; petioles 2–6 mm. long, appressed pubescent. Stipules subulate, ± 1 mm. long. Flowers borne in terminal and lateral nearly or quite leafless much-branched secund vaguely panicle-like inflorescences 1–2 dm. long, each node bearing on the upper side a glomerule of 3–10 flowers, the curved branches densely pubescent with appressed brownish hairs ± 0·1 mm. long; pedicels 1–2 mm. long in flower, 3–4 mm. in fruit, densely pubescent. Cup and backs of sepals densely pubescent with appressed brownish hairs ± 0·1 mm. long. Sepals deltoid, 1–1·3 mm. long. Petals ± 1 mm. long, strongly reflexed, glabrous. Disk densely pubescent with appressed yellowish hairs. Fruit 5–7 cm. long, the wing 8–10 mm. broad, minutely pubescent with appressed hairs.

UGANDA. Mengo District: Mabira Forest, Mulange, Nov. 1922, *Dummer* 5605!
KENYA. N. Kavirondo District: Kakamega Forest, 15 Oct. 1953, *Drummond & Hemsley* 4793!
TANGANYIKA. Mbulu District: Lake Manyara National Park, Bagoyo R., 17 Feb. 1964, *Greenway & Hunter* 11190! & 14 Nov. 1963, *Greenway & Kirrika* 11018!
DISTR. **U4**; **K5**; **T2**; Nigeria, S. Tomé, Cameroun, Congo
HAB. Rain-forest and riverine forest; 990–1600 m.

SYN. *Celastrus diffusus* G. Don, Gen. Syst. 2: 6 (1832)

11. MAESOPSIS

Engl., P.O.A. C: 255 (1895)

Trees. Leaves usually subopposite, varying from somewhat alternate to strictly opposite, petiolate; blades penninerved, strongly glandular-serrulate, the tertiary nervature rather elegantly percurrent, each juncture of a secondary nerve and the midrib marked beneath by a peculiar minute (? glandular) process. Stipules subulate, caducous. Flowers perfect, 5-merous, in divaricately branched axillary cymes often several cm. long. Sepals deltoid. Petals not at all clawed, extremely concavo-convex so that the anther is almost totally obscured even at the time of its dehiscence. Anther sessile or nearly so. Disk thin, lining the cup at its rim, minutely 10-lobed with 2 lobes opposite each anther and petal. Ovary elongate, apparently 1-celled, the mesocarpous tissue somewhat spongy or fleshy and with numerous (?

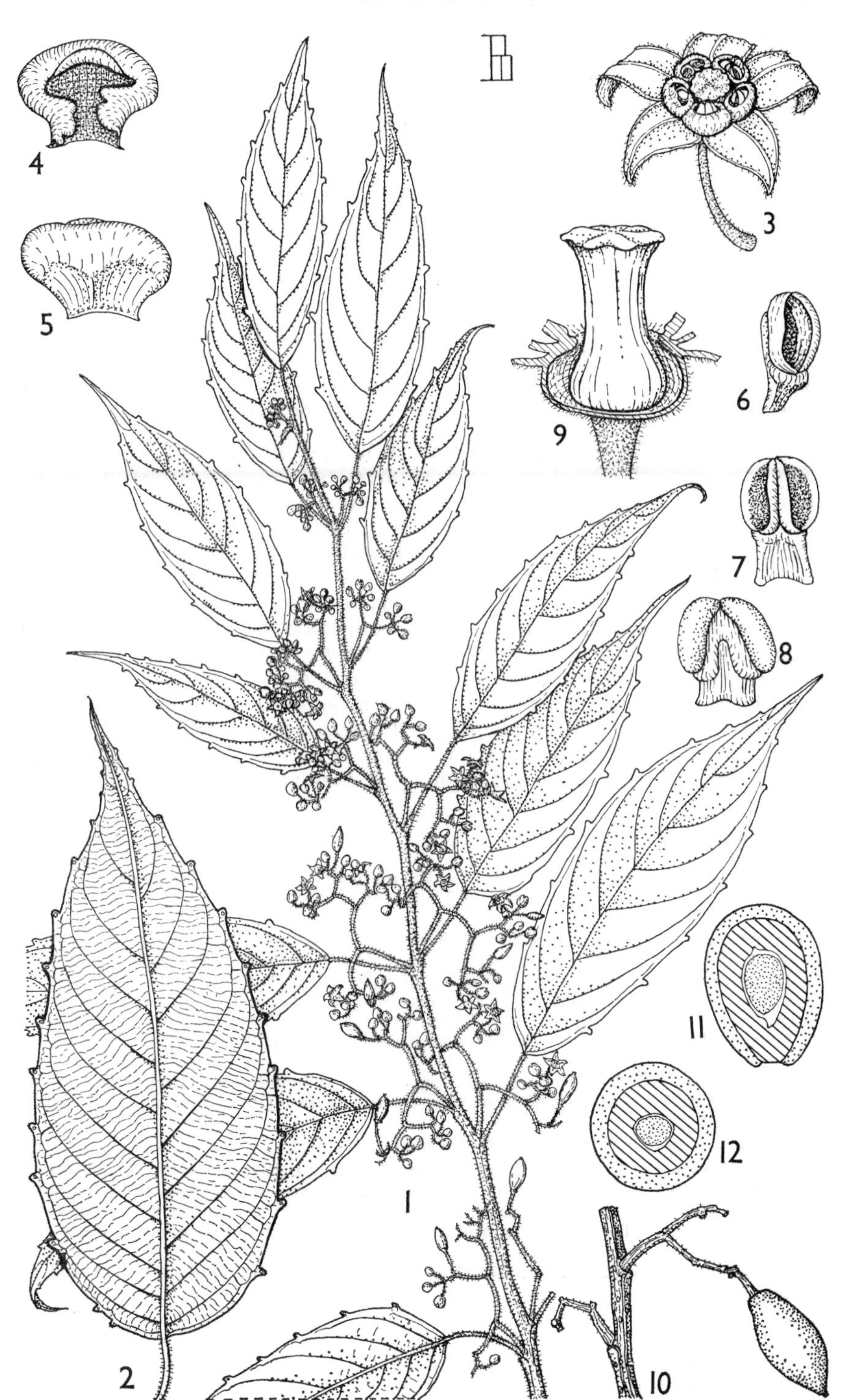

FIG. 12. *MAESOPSIS EMINII* subsp. *EMINII*—**1,** flowering branch, × $\frac{2}{3}$; **2,** leaf, showing undersurface, × $\frac{2}{3}$; **3,** flower, × 10; **4, 5,** petal, from inner and outer side respectively, × 20; **6, 7, 8,** stamen, side, front and back view respectively, × 20; **9,** flower with parts cut away to show gynoecium, × 20; **10,** portion of fruiting branch, × $\frac{2}{3}$; **11, 12,** longitudinal and transverse sections of fruit respectively, × 1. 1, from *Eggeling* 108; 2, from *Conrads* 424; 3–9, from *Dawe* 198; 10, from *C.M. Harris* in *F.D.* 408; 11, 12, from *Styles* 51.

resin) cavities, the endocarp thin and becoming stony; style projecting only slightly above the petals, simple, expanded at top into a narrowly mushroom-shaped form and usually marked at top by furrows into (4)5 portions, each of which bears 2 minute tooth-like lobes. Fruit a drupe with a single 1-seeded stone.

A monotypic tropical African genus. The ovary and especially the style distinguish this from all other *Rhamnaceae*. If the flowers are truly protogynous as reported by N. Hallé, Fl. Gabon 4: 52 (1962), this would also distinguish them from all other rhamnaceous flowers. The genus perhaps deserves a subfamily to itself.

NOTE. *Maesopsis stuhlmannii* Engl. is apparently a synonym of *Macaranga monandra* Muell. Arg., *Euphorbiaceae* (T.T.C.L.: 468 (1949)).

M. eminii *Engl.*, P.O.A. C: 255 (1895); Eggeling & Harris, Fifteen Uganda Timbers: 99 (1939); T.T.C.L.: 468 (1949); I.T.U., ed. 2: 323, photo. 52, fig. 68 (1952); Verdc. in B.J.B.B. 27: 352 (1957); K.T.S.: 388, fig. 76 (1961). Type: Tanganyika, Bukoba, *Stuhlmann* 971 (B, holo. †, M, iso. fragment!)

Trees (5–)15–25(–42) m. tall; trunk (1–)2–5(–10) dm. thick; bark silvery grey with vertical twisted furrowing; slash red outside, yellow near the wood; heartwood yellowish, darkening on exposure to reddish brown. Year-old branchlets glabrescent, smooth, brownish, lenticellate; youngest branches dark, puberulent to nearly glabrous. Leaf-blades ovate-elliptic to oblong-ovate, 7–14 cm. long, 2·5–6 cm. wide, lustrous above, paler beneath, glabrous except when quite young, at base rounded to subcordate, acuminate, at margins with rounded ± salient teeth or projections 0·3–5 mm. long, on each side of midrib with (6–)7–10 secondary nerves; petioles 6–12 mm. long, puberulent to glabrescent. Stipules 2–6 mm. long, puberulent. Cymes 1–5 cm. long, many-flowered; primary peduncle 4–25 mm. long; ultimate pedicels 1–3(–6) mm. long. Sepals ± 1·5 mm. long. Drupe (only 1, 2 or rarely 3 fruits set per inflorescence) obovoid or narrowly so, 22–30 mm. long, 10–16 mm. thick, the style and stigma persistent; outer portion of mesocarp fleshy.

subsp. **eminii**

Rather large trees on the average. Leaves on drying becoming olive-green in colour; marginal teeth rather long, usually ± 2 mm. long. Fig. 12, p. 37.

UGANDA. Bunyoro District: Budongo Forest, Feb. 1932, *C. M. Harris* in *F.D.* 406!, 407! & 408!; Busoga District: Lolui I., 11 May 1964, *G. Jackson* U44!; Mengo District: Mabira Forest near Najembe, 14 Apr. 1950, *Dawkins* 569!
KENYA. N. Kavirondo District: Kakamega Forest, *Machin* 864!
TANGANYIKA. Biharamulo District: Ussui, 19 June 1913, *Braun* in *Herb. Amani* 5537!; Mwanza District: Ukerewe I., *Conrads* 5313!; Kigoma District: Uvinza, Nov. 1954, *Procter* 303!
DISTR. **U**2–4; **K**5; **T**1, 4; Zambia, Congo, Angola; populations from Congo W. to Liberia have been distinguished as subsp. *berchemioïdes* (Pierre) N. Hallé, Fl. Gabon 4: 51 (1962)
HAB. Rain-forest and riverine forest; 800–1200 m.

NOTE. The records for **K**4 and **T**3 are apparently all of cultivated trees. This is one of the more important timber trees in East Africa. Its ease of propagation, high quality lumber and fairly good resistance to fungi and insects encourage its use in other tropical areas. The typical subspecies is now cultivated in Sumatra, Java and Borneo.
Verdcourt in B.J.B.B. 27: 352 (1957) mentions that the fruits are dispersed by hornbills.

INDEX TO RHAMNACEAE

GEOGRAPHICAL DIVISIONS OF THE FLORA

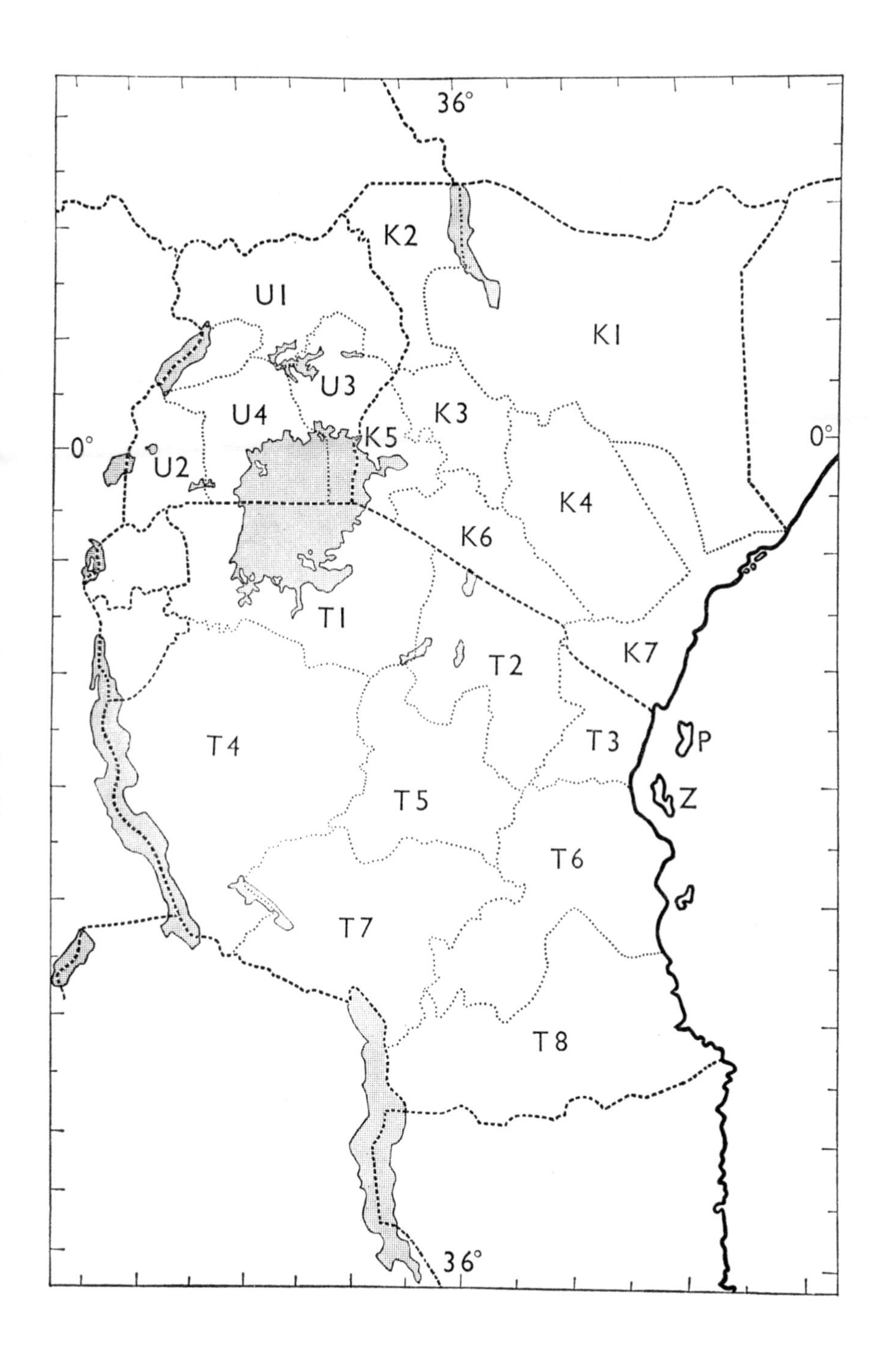